100 FIRST-PRIZE MAKE-IT-YOURSELF SCIENCE FAIR PROJECTS

Written & Illustrated by
Glen Vecchione

GOODWILL PUBLISHING HOUSE
B-3, RATTAN JYOTI, 18, RAJENDRA PLACE
NEW DELHI-110008 (INDIA)

Edited and page layouts designed by Jeanette Green

© Glen Vecchione

All rights reserved.

This special low priced Indian reprint is published by arrangement with Sterling Publishing Company, Inc., New York, USA.

No part of this publication may be reproduced, stored in a retrieval system, or transmitted in any form or by any means, electronic, mechanical, photocopying, recording or otherwise, without the prior written permission of the publisher.

Price : in USA – US $ 13.95
         in India – Rs. 99.00

Published in India by
GOODWILL PUBLISHING HOUSE
B-3, Rattan Jyoti, 18 Rajendra Place
New Delhi-110008 (INDIA)
Tel. : 5750801, 5755519
Fax : 91-11-5763428

Printed at
Kumar Offset
New Delhi

Contents

Introduction	5
Up, Down, All Around	9
Bird Feathers & Bug Boxes	42
Creative Concoctions	60
Find Out about Food	89
Excellent Electrics	111
Natural Laws	134
Puzzling Plant Projects	162
Fly, Float & Sink	181
Metric Equivalents	201
Index	202

Acknowledgments

I would like to thank these people for their help building and testing the projects in this book:

Holly, Rick, and R. J. Andrews
Lenny, Claire, and Kyrstin Gemar
Cameron and Kyle Eck
Lewis, Hava, and Tasha Hall
Jeri, Bryan, and Jesse James Smith
Tony and Kasandra Ramirez
Joe, Kate, and Micaela Vidales
Debbie and Mark Wankier
Stephen Sturk
Nina Zottoli
Eric Byron
Andy Pawlowski

A special thanks to my friend David Lee Ahern

And as always,
For Joshua, Irene, and Briana Vecchione

Introduction

Your Project & the Scientific Method

This how-to book of 100 science-fair projects will teach you, step-by-step, how to create an exciting project that not only demonstrates good scientific practice but gives you the first-prize edge. Collected into eight chapters, each project briefly examines a scientific topic, such as plant behavior or planetary motion. In turn, each topic suggests questions that the completed project will answer.

But selecting a science-fair project means more than just following instructions about how to build it. Today, science-fair competitions require that you research, question, prepare, test, and draw conclusions about your project. This procedure, called the *scientific method*, is as important as the project itself. How carefully you follow the scientific method in preparing your project will be an important consideration when the judges look at it.

So, what is the scientific method? Over the centuries, the scientific method has become the most important procedure for scientific research, discovery, and invention. Simply described, the scientific method researches a topic, asks a question, and then tests a potential solution, or *hypothesis*, by performing an experiment. The experiment, often repeated many times, yields information that either supports or contradicts the hypothesis. Following this procedure yields information that will add to your general understanding of a topic.

In this case, the "experiment" is your science-fair project, and you only have to create it once. Our 100 projects have been organized to help you find the scientific topic that most interests you—part of the research stage of your project-making. A short introduction to each project will help raise questions in your mind, followed by step-by-step instructions for creating your project. Finally, our results and explanation sections will help you understand what your project demonstrates and how this information can answer broader questions about science.

Although you'll find many projects that require model-constructing, other projects require you to collect data with the emphasis on analyzing and displaying your results attractively. This means that we've included something for everyone, from bridge-building to charting tree-slope data; from constructing a shadow box that demonstrates moon phases to figuring the per-inch air pressure against a plunger. And some projects, like those explaining stellar parallax, even require a little game-playing activity by participants. It all adds up to fun, and having fun with science is a great way to learn about science.

All projects are safe, and all use inexpensive and easily found materials. When appropriate, we suggest alternate materials for variations of projects. We also provide tips on how to display each project, and most projects are followed with an interesting additional story or facts that will encourage you to learn more about what you're doing. All projects have been built, tested, and redesigned when appropriate to make their results more pronounced or dramatic.

Of course, the ultimate goal of a book like ours is to encourage you to do a little experimenting on your own. Redesign a project and tell us about it. Let us know how you do in your science-fair competition and how many prizes you bring home. We hope you'll be a first-prize winner!

Science-Fair Rules

On the local level, rules for science fairs vary, although they become more standard for regional competitions. Here are some common rules for most science fairs.

Introduction

The student (not a teacher or other adviser) chooses the project.
Only one project per student.
The student can ask for help with research, creating the project and writing the report, but she must distinguish between her own work and the work of her helpers.
All projects must be set up and taken down during designated hours, determined by the judges.
Students must be available to speak to the judges at an appointed time.
Students must take all responsibility for the safety of their exhibited projects.

These additional rules may apply to a school or school-sponsored science fair.

No live cultures of bacteria or fungi.
No animal or human body parts.
No live vertebrates—animals with backbones such as fish, frogs, birds, or mammals.
No dangerous chemicals that can easily ignite or give off toxic fumes.
Electrical projects must use 110 volts and are limited to 500 watts.
All projects must follow county, state, and federal laws and regulations regarding wiring, toxic materials, fire hazards, and structural safety.

Ask your teacher if you're unsure about whether your project follows the rules. For example, some schools allow live cultures, and some allow live vertebrates provided the animals are not harmed in any way. Insect projects, as long as the insects remain safely contained, are allowed at most science fairs.

Getting Started

Remember: choosing a project is only the beginning. The topic must interest you, too. Creating your project is just one phase of the procedure you'll use to learn more about your topic. To help get you started, let's break this approach into six steps.

1. Choose a project.
2. Research the topic.
3. Ask a question.
4. State a hypothesis.
5. Construct a project to test the hypothesis.
6. Draw a conclusion.

First, read some introductory sections for projects in this book (step 1). If a project interests you, find additional sources where you can learn more about the topic (step 2). For example, if you're interested in one of the astronomy projects, go to the library and learn a little more about astronomy. You'll probably find that you have lots of questions about astronomy. Write them down (step 3). Now, could one of your questions be answered by building the astronomy project? But before you build your project, think about how you might be able to answer your question by writing a hypothesis (step 4). Next, build your astronomy project to test your hypothesis (step 5). Finally, find the answer to your question by noting whether the results of your project support or contradict your hypothesis (step 6).

Keeping a Project Journal

At every step in creating your project, make notes in a small, spiral-bound notebook. These notes should record why a particular topic and project interested you. Also record where you found project materials, the cost of building the project, whatever difficulties you had, and when you needed help. Don't worry about taking perfect notes that you have to show to anyone. Your notes will help you remember the details of the project when it's time to write your report.

Writing a Project Report

Your report is the finished document you submit to the judges, and it should be a carefully written record of your project from start to finish. The notes in your journal will provide most of the material in your report, but make sure you write the report so that someone unfamiliar with the topic or with what your project demonstrates can clearly understand it. Although most science fairs don't require a standard report format, a good way to organize report material is to divide it into seven short sections: (1) title page, (2) contents, (3) abstract, (4) introduction, (5) building the project, (6) conclusion, and the optional (7) extra section.

Title Page Rules vary for title pages, but usually the name of the project should appear boldly in the center of the page. *Leave your name off the title page for judging. You may be assigned a number or other code.*

Contents On the contents page, list the sections of your report and give the page number for each section. This will help judges turn easily to a particular section if they want more information about your project.

Abstract The abstract is a brief overview (one or two paragraphs) of your selected topic and how your project fits into that topic. Include any questions you have about your topic as well as your hypothesis. Remember, a hypothesis is the untested conclusion you formulate that will either be proved or disproved by your project. The abstract should also describe the procedure you followed when creating your project and what you hoped to learn from it.

Introduction In the introduction, explain what interested you about this project and any experiences in your background that may have led you to choose this particular project rather than another. What does this project mean to you?

Building the Project In this section, describe in detail the experience of creating your project, whether model-building, calculating, measuring, or collecting data. Talk about collecting or buying materials for the project. Mention the tasks you could do alone and those that required adult help. Don't be afraid to be critical of your project. Clearly state what you enjoyed or did not enjoy and what you found easy or difficult about your project.

Conclusion The conclusion states what you learned from the project and whether or not that information supports or contradicts your hypothesis. You may find the results of your project-making inconclusive. If this is the case, try to find out why. Were the instructions followed carefully? Were materials substituted? Were measurements rechecked? Many scientists consider themselves lucky if an experiment yields accurate results the first time. If you can repeat your project without too much trouble, do it, and see if you get more conclusive results the second time.

Extra Section As a finishing touch to your report, you can also include a section that lists sources of additional information you used, such as books, brochures, CD-ROMs, or videos. Also in this section, you can give credit to people who helped you with your project, like your friend or your older brother—even your mom or dad. Closing your report with a section like this gives it a thorough and professional quality.

Tips for Your Display

Even the most exciting project won't attract attention if it isn't attractively displayed. Taking time to stage your project carefully can make all the difference in whether the judges stop to read your material or only glance at it and hurry to the next display.

All science fairs require that you display your project on a table and backboard. A backboard consists of a wooden or cardboard stand that folds at the sides so that it rests on the table unsupported. You can either make a backboard or purchase one in any science or hobby shop. The allowable size and shape of the backboard can differ between science fairs, so check with your teacher before you build or buy one.

As for organizing materials for the display, remember: you want to show that your project is only part of an exploration into a scientific topic, an exploration that used the scientific method. This means that you need to clearly attach your abstract, hypothesis, and conclusion to the backboard, as well as any photographs you took when creating your project. If your project consists of a model, the model should sit in the center of the table so that it can be seen without blocking anything else. At the top center of your backboard, clearly print the name of your project.

The key to a successful display is to avoid crowding everything together—descriptions, photographs, models. You want observers to be drawn into your project and learn more about it. Let your display tell a story about your project, interests, and ideas. It should welcome anyone who comes to look. It should say something about you—your creativity, hard work, careful thinking, and love of science.

About the Metric Equivalents in This Book

The metric equivalents used in parentheses in these science-fair projects have been rounded off for convenience. So, we'll say 450 grams or even 500 grams to make a project requiring 1 pound of material, aware that the metric equivalent of 1 pound is 454 grams. And most measurements in this book, whether dry or liquid, assume capacity (liquid) measures in convenient milliliter conversions from teaspoons, tablespoons, and cups, etc.

Weights of these dry or liquid substances measured in teaspoons, cups, or quarts, etc., of course, will vary.

Consult the Metric Equivalents table on p. 202, or ask your math teacher for help.

Up, Down, All Around

Earth's Magnetic Field
Dipping Compass
Cyanometer
Skywatching
Moon Box
Twinkling Starlight
Stellar Refraction
Sheet Erosion
Split a Stone
The Parallax Principle
Parallax Shift
Height Calculation
Rare-Earth Metals
Magnetic Filtration
Box Periscope
Mapping the Ocean Floor

Earth's Magnetic Field

You Will Need

- Bar magnet
- Iron filings (or a steel nail and a file)
- Old pepper shaker
- Coffee-can lid
- 2 sheets of stiff white paper
- Spray bottle
- White vinegar
- Ruler
- Marking pen

Earth Map with Bar Magnet Beneath
To Illustrate the Earth's Magnetic Fields

Large or small, all magnetic fields have a similar shape. The Earth's gigantic magnetic field, wrapping from the North to South Pole, looks very much like the simple magnetic field that surrounds a bar magnet. You can see this for yourself in this project.

Procedure

1. Use the coffee-can lid to trace a circle onto one of the sheets of paper. Draw continents inside the circle, turning it into a simple map of the Earth.
2. Make two narrow folds at opposite sides of your Earth map so that the map sits just barely above the bar magnet, which you'll place underneath.
3. Position the magnet so that its poles line up with the north and south poles of your Earth map.
4. Fold the second sheet of paper into a funnel and place one end into an old pepper shaker.
5. Pour iron filings into an old pepper shaker. If you can't find iron filings, you can make them by scraping an iron bar or steel nail against a file. Collect enough filings to thinly coat a piece of paper, then fold the paper and gently tap the filings into the pepper shaker.
6. Carefully sprinkle iron filings over the map. Gently blow on the filings to spread them across the page.
7. Fill the spray bottle with white vinegar and carefully spray the map. Make sure you hold the bottle far enough away from the filings so that you don't disturb their positions. Allow the vinegar to dry overnight, then brush the filings from the map.

Result: As you sprinkled the filings over the map, something amazing appeared—the magnetic field of the bar magnet, traced in filings. This magnetic field accurately recreated the magnetic field of the Earth. The vinegar caused the filings to rust, leaving a clear imprint of the magnetic field on the paper.

Explanation

Magnetic lines of force come together at two points, or magnetic poles. Although scientists have long searched for exceptions, every magnet known to man has a north and south pole that can never be separated. So, large or small, all magnetic fields resemble each other.

Display Tip

Document your experiment with photographs. Exhibit the finished map next to the photographs. Research and diagram several shapes of magnetic fields and identify their poles.

Did You Know?
By studying the position of iron and magnetite particles in ancient clay banks, scientists can figure out where the Earth's magnetic poles were thousands of years ago. Like tiny compass needles frozen in time, the particles point to a north magnetic pole that no longer exists, close to what we now call the geographic South Pole! So, many scientists now believe that the Earth's magnetic poles were once completely reversed.

Dipping Compass

You Will Need

- Wire hanger
- Wire clippers
- Small Styrofoam ball
- Knitting needle
- Compass
- 2 glass tumblers, same size
- Protractor
- Wood block
- Strong bar magnet

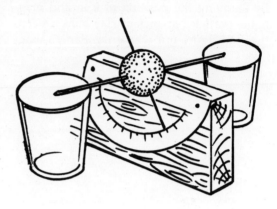

If you removed the needle from the face of a compass and tied it to a piece of string, the needle would not only point north but dip along the lines of the Earth's magnetic field. Scientists have another name for this kind of dipping compass: *inclinometer*. You can trace the lines of the Earth's magnetic field by constructing an inclinometer.

Procedure

1. Have an adult help you clip a straight piece of wire from the wire hanger with the wire clippers.
2. Push the piece of wire through the Styrofoam ball so that equal lengths of wire stick out from opposite sides of the ball.
3. Stick the knitting needle through the ball at an angle perpendicular to the piece of wire hanger.
4. Place the whole compass assembly so that the ends of the knitting needle rest on the tumblers and the Styrofoam ball and hanger sits between the tumblers.
5. Swivel the hanger into a horizontal position. Using the compass as a guide, turn the tumblers so that the hanger points north–south.
6. For about a minute, stroke the north end of hanger with the north pole of the bar magnet. This magnetizes the hanger and turns it into a north-seeking compass needle.
7. Attach the protractor, upside down, to the side of the wood block. Slide the wood block between the tumblers so that it sits just beside the Styrofoam ball.
8. Allow the needle to settle into position and observe its angle.

Result: The north end of the needle slowly dips between the tumblers to about a 45° angle.

Explanation
The inclinometer's angle of dip reflects the lines of magnetic force at your latitude. At the Earth's Equator, the needle would be completely horizontal; at the North Pole, the needle would stand upright. For most northern latitudes, the needle will point at about a 45° angle.

Display Tip
Document the construction of your instrument with photographs, and display the finished model. Make a chart of the Earth's magnetic lines of force and show how the inclinometer might dip at various latitudes.

Did You Know?
Since like poles of a magnet repel and opposite poles attract, how can a north-seeking

magnetized needle point to the Earth's *North* Pole? The answer is that the magnetic *south pole* lies near the Earth's geographic North Pole, and the magnetic *north pole* lies near the Earth's geographic South Pole. This topsy-turvy arrangement has proved so confusing over the years that scientists now prefer to call the Earth's magnetic poles the north- or south-seeking poles.

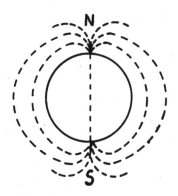

Earth's Magnetic Field

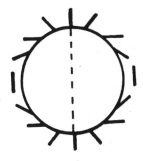

Positions of a Dipping Compass

Cyanometer

You Will Need

- Cardboard box lid
- 2 sheets of 8½ × 11-inch white paper
- Black, blue, white, brown, and red poster paint
- Paintbrush
- 16 small paper cups
- 16 rubber bands
- Cellophane
- Pencil compass
- Navigator's compass
- Ruler
- Marking pen
- Craft knife
- Protractor
- Red string
- Rubber cement
- Metal nut
- Pushpin
- Brass fastener

If you look up on a cloudless day, you'll notice that one part of the sky seems a darker blue than the rest. This deep-sky area moves around depending on the hour of day and position of the Sun. Astronomers keep track of the changing hues of the sky with the aid of a simple instrument called a *cyanometer*.

The term *cyanometer* is based on the word *cyan*, which means the color blue mixed with yellow.

You can combine this project with the *Skywatching* project.

Part 1 Drawing the Chart

Procedure

1. Draw a large circle on the first sheet of paper using the pencil compass.

2. Draw three smaller circles inside.

3. Use the ruler to divide the circle into twelve pie slices, but keep the center circle empty. These circles represent the entire visible sky, from horizon to directly overhead.

4. Where one of the spokes intersects the largest circle, write 0 degrees (0°). Move into the next circle and write 30 degrees (30°), then 60 degrees (60°), and 90 degrees (90°) in the center circle. These degrees represent *angular altitude* in the sky, with 0° for the horizon and 90° for directly overhead.

5. Draw the four cardinal compass points outside the circle, and number the sections inside the circle 1–12, following the diagram.

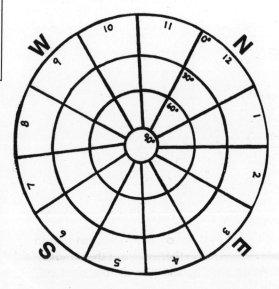

6. Make at least ten photocopies of your chart.

Part 2 Patches of Blue

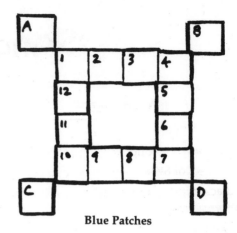

Blue Patches

Procedure

1. Draw sixteen 1-inch (2.5-cm) squares on the second sheet of paper.
2. Take twelve of the sixteen paper cups and mix blue, white, and black paint into twelve shades of blue. The shades should go from light blue to deep (but not dark) blue.
3. Paint each of the first twelve squares a slightly darker shade of blue.
4. Use the marker to number your painted squares from 1 to 12, with *1* for the lightest shade and *12* for the darkest.
5. In the four remaining paper cups, mix a light shade of blue. In the first cup, add 3 drops of brown paint. In the second cup, add 6 drops of brown paint. In the third cup add 3 drops of red paint, and in the fourth cup add 6 drops of red paint.
6. Paint four squares in these colors and letter them A, B, C, and D, with letters A and B containing the blue-and-brown paint mixture.
7. Cover the paint that remains in each paper cup with a piece of cellophane wrapped tightly with a rubber band. Label all the cups.

ed) and stick a pushpin through the center hole.

7. Cut a 6-inch (15-cm) piece of red string, and tie one end around the metal nut. Tie the other end around the pushpin.
8. Glue the compass to the bottom edge of the box.
9. Take one of your photocopied charts, and attach it to the bottom part of the box with a brass fastener. Make sure the chart rotates easily. You now have an instrument that will allow you to take accurate measurements of shades of blue in the sky.

Part 3 Viewing Box

Procedure

1. Turn the box lid upside down with the underside facing you.
2. Draw a 2-inch (5-cm) window in the center of the box lid, near the top.
3. Have an adult help you cut out the window with the craft knife.
4. Paint the inside of the box lid black. Allow the box to dry.
5. Cut out all the color squares painted in part two, and glue them around the window, following the top diagram in the next column.
6. Glue the protractor to the left edge of the box lid (or right edge, if you're left-hand-

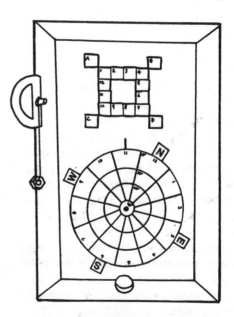

Finished Cyanometer

Skywatching

You Will Need
- Cyanometer
- Charts from Cyanometer project

Here you'll get a chance to use your cyanometer (from the Cyanometer project) to take accurate blueness measurements of the sky. Record your results in a color chart to display for the judges.

Procedure

1. Using the cyanometer's compass to guide you, stand facing north, and turn the chart so that the N is at the top.
2. Look through the window and make your first sky observation at the horizon. Notice how the thread hanging against the protractor indicates 0°.
3. Match the color of sky against one of your color patches, and write the number (or letter) on the chart.
4. Imagine yourself standing in the center of a big clock. Turn to face one o'clock and rotate the chart so that the "1" is at the top.
5. Make your second observation, matching the sky with a color patch and recording the number or letter.
6. Face two o'clock, rotate the chart, and make your third observation.
7. Continue this procedure until you've made a complete circle and observed the entire horizon.
8. Tilt the box upward to 30° (use the string and protractor to guide you), and repeat steps 2–7. Continue this procedure until you've covered the entire sky.
9. Here's where your leftover paint comes in handy. Remove your chart from the viewing box and color it in, using the numbers and letters to guide you. You'll wind up with a complete painted map of the sky. Make sure that you indicate the date and time of day on each painted chart.
10. Repeat steps 1–8 several times a day in order to compare charts.

Result: You'll be surprised to notice many shades of blue in an ordinary sky. But the deep-sky region occurs in a different place, depending on the time of day you make your observation.

Explanation

The numbered swatches of blue represent the clear-sky variations of blue (including deep sky blue), while the lettered swatches show "blue pollution" by smog or reddening sunlight. The later in the day you make your observations, the more brown you'll see at the horizon and the more red you'll notice toward the western region of the sky.

But why is the sky blue in the first place? White sunlight actually consists of many different colors, or wavelengths, of light mixed together. When sunlight passes through the Earth's atmosphere, molecules of air and particles of dust absorb most of the light but scatter the rest. The size of the particles and thickness of the atmosphere determine what kind of scattering occurs and the resulting color of the sky. Smaller particles scatter shorter wavelengths (blue light), and larger particles scatter longer wavelengths (red light). Scientists call this *selective scattering*.

A deep blue sky not only means that the particles in our atmosphere are very small, but also that the air is relatively free of dusts, water vapor, or man-made pollutants. These impurities will "wash out" the deepest hues of the sky.

As for the deep-sky region, or area of darkest blue, the position of the Sun has everything to do with where you'll find it. This is because the Sun's light refracts (bends) as it travels through the atmosphere, and this bending affects the color of the sky.

Here's a simple formula for finding the deep-sky region at any time of day. Imagine the Sun traveling in a huge half-circle across the sky as if attached to a giant protractor. When the Sun is low on the horizon, the darkest blue occurs about 90° away—toward the innermost circle of your cyanometer. When the Sun is directly overhead or at the center of your cyanometer, you'll find the deepest blue at about 45° in all directions, like a halo. By careful observation at different times of day, you should be able to determine additional angular relationships between deep sky and the position of the Sun.

Display Tip

For an interesting exhibit, display your painted charts showing deep sky at various times of day. Mount your paintings carefully, even framing them if possible.

Did You Know?

Earth is probably unique in our solar system for having a blue sky. Photographs sent back from the surface of Mars show a featureless pink sky. Astronomers believe that the pink color comes from the presence of iron oxides in the dust particles that continually blow around the planet.

Moon Box

You Will Need

- Shoe box with lid
- Styrofoam ball (about 1 inch or 2.5 cm in diameter)
- Black thread
- Paper clip
- Brass fastener
- Rubber cement
- Blue, white, black, and yellow poster paint
- Black construction paper
- Pencil compass
- Marking pen
- Ruler
- Scissors
- Small flashlight
- Electrical tape

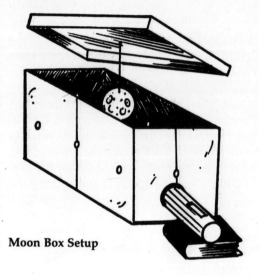

Moon Box Setup

Right off the top of your head, do you know what the Moon will look like tonight? Although most people can recognize and name the different phases of the Moon, it usually takes something like this project to help you understand them.

Part 1 Punching the Peep Holes

Procedure

1. Remove the lid of the shoe box. Use the needle of the pencil compass to punch eight holes around the middle of the box. Enlarge each hole by pushing the pencil through it, and then write a number above each hole.

2. Use the ruler and pencil to draw four lines from the top to the bottom of the box that cross over holes 1, 3, 5, and 7. These lines divide your box into four lunar quarters.

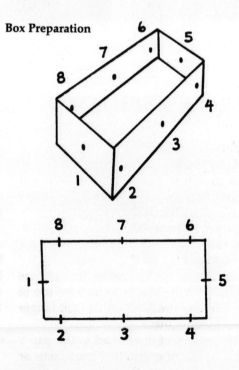

Box Preparation

3. Paint the areas between holes 1–3 and 5–7 dark blue, and paint the areas between

holes 3–5 and 7–1 light blue. Paint the outside of the box lid dark blue.

4. Paint the inside of the box and lid black. Allow the paint to dry.

5. Put the lid on the box. Use the marking pen to write the numbers 1–8 on the side of the lid, over the holes.

6. About 1 inch (2.5 cm) below hole #5, cut another hole, just large enough to fit the front part of the flashlight.

7. Use the pencil compass to draw eight small circles on the black construction paper.

8. Use the yellow poster paint to reproduce the phases of the Moon shown in the diagram, and glue each circle above its matching hole on the box.

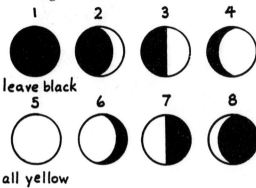

Phases of the Moon

Part 2   A Model Moon

Procedure

1. Make "craters" in the Styrofoam ball by pushing the eraser end of the pencil into the Styrofoam. You can also paint the Styrofoam to create a more Moonlike texture.

2. Unbend a paper clip and push it into the Moon so that you have a hook on which to tie the thread.

3. Use the ruler to measure the distance between one of the box holes and the top of the lid. Cut the thread about an inch longer than this measurement.

4. Tie one end of the thread to the paper clip, and tie the other end to a brass fastener that you push through the top of the box lid at the center.

5. Carefully put the lid (with hanging Moon) on the shoe box.

6. Prop the flashlight up on some books so that the front part fits through the hole in the box you cut earlier. Put electrical tape around the sides of the flashlight where it joins the box.

7. Switch on the flashlight and look through the peepholes.

Research: Each hole simulates a different phase of the Moon due to the changing angle of the flashlight and Styrofoam ball.

Explanation

When astronomers speak of the phases of the Moon, they mean how much of the sunlit part of the Moon we can see from Earth. Each of the eight peepholes shows a different view of the Moon as it moves through the lunar month. Numbers 1, 3, 5, and 7 represent what astronomers call *cardinal phases* of the Moon, because each phase has a distinct shape and occurs at a definite moment in time. Numbers 2, 4, 6, and 8 represent *stretches*, since the Moon is continually growing or shrinking during this time between cardinal phases.

The cardinal phases of the Moon have the names *new moon*, *half-moon* (also called *first* and *last quarter moon*), and *full moon*. You might know the stretches of the Moon as *waxing* (growing) *crescent*, *waxing gibbous*, *waning* (shrinking) *gibbous*, and *waning crescent*. The waning crescent Moon is also sometimes called *descrescent*.

Although the new moon (represented by the black circle) has a peephole, the new moon is actually invisible to us. During the new-moon phase, the Moon sits between the Earth and the Sun, so we can't see the sunlight reflecting off it. But you can see a new Moon sometimes, and it's a rare and amazing sight. Can you guess when this happens? Look through peephole #1 for a good clue. You'll see the light peeking out from the edges of your model Moon. When the real Moon passes in front of the Sun, it creates one of the most spectacular and lovely sights in nature: the solar eclipse.

Display Tip

You certainly want to keep your model on display so that the judges can peek through the holes and experience the phases of the Moon for themselves. You should also document how you constructed your model through photographs. Round out your display with additional information about the Moon. You can obtain a wonderful information kit, along with beautiful photographs of the historic 1969 Pioneer Moon landing, by writing NASA in Houston, Texas.

Did You Know?

You might wonder why astronomers gave the name first and last quarter moon for something as easy to remember as half-moon. But if you think of the Moon as a sphere, and you visualize what part of the sphere reflects the sunlight, you can see that what appears as a half-moon from Earth is actually only a quarter of the Moon's sphere. This means that you could also call the full moon a half-moon if you wanted ... but nobody seems to like that idea.

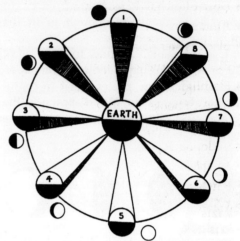

Direction of Sunlight

Twinkling Starlight

You Will Need

- Deep glass bowl (transparent)
- Aluminum foil
- Black poster paint
- Black construction paper
- Flashlight
- Stack of books
- Small and large nails
- Cellophane tape
- Marbles
- Paintbrush
- Water

Star Bowl

Why does the light from distant stars appear to twinkle when you look up at the night sky? This project will help you find out by creating light refraction through turbulent water.

Part 1 Constructing the Star Bowl

Procedure

1. Cover the bowl in aluminum foil so that you can see the shiny side of the foil through the glass.
2. Fill the bowl three-quarters full with water.
3. Make a hole in the foil large enough for the head of a flashlight.
4. Prop the flashlight up on a few books, and place the light so that it shines through the bowl and into the water. Wrap some more foil around the side of the flashlight to keep light from leaking out.
5. Cover the top of the bowl with another sheet of aluminum foil, shiny side down. Paint this sheet of foil black.
6. Use the small nail to punch holes through the painted foil in a constellation pattern. Use the larger nail to make a few bigger holes for planets.
7. Tear away a small piece of foil at the edge of the bowl, just large enough for a marble, and cover the hole with a small piece of black construction paper. Tape the construction paper at one edge so that it makes a kind of flap.

Part 2 Twinkling Star Show

Procedure

1. Take your star bowl to a dark place, or have a blanket that viewers can place over themselves while observing the "stars."
2. Switch on the flashlight and look down at the top piece of black foil. Notice the steady light coming through all the holes.
3. Lift the flap and drop several marbles into the bowl. Enjoy the effect of twinkling stars.

Result: The water turbulence from the dropping marbles causes the smaller holes, or "stars," to twinkle. The light from the "planets" twinkles also, but the effect is much less noticeable.

Explanation

Dropping marbles into the bowl causes the water in the bowl to ripple. This rippling bends and distorts the light as it travels through the water, reflected from the aluminum foil. Stars appear to twinkle because their far-away light passes through millions of space miles as well as through the different layers of air in the Earth's atmosphere. These layers have different thickness and are sometimes turbulent, so starlight does not travel in a straight line, but rather bends or *refracts* thousands of times per second (we say it twinkles) before it reaches the Earth. The stronger light from the much closer planets is much less affected by the turbulence of the Earth's atmosphere.

Display Tip
Document the construction of your star bowl and display the finished model. Have a blanket or dark piece of cloth handy so that viewers can experience the twinkling to its full effect.

Did You Know?
Astronomers measure the distance of stars in *light-years*, or how far light travels in one year. Since the speed of light is over 186,000 miles per second or nearly 300,000 kilometers per second, you can imagine some pretty big numbers when light-years are translated into miles. For example, the distance light travels in one year (light-year) is 5.88 trillion miles or 9.46 trillion kilometers.

But just how far away from Earth is Proxima Centauri, the closest star? If the Earth were shrunk to the size of a grain of sand and placed in the city of San Francisco, you'd find Proxima Centauri in Arizona, just south of the Grand Canyon.

Stellar Refraction

You Will Need

- Black poster board
- Black plastic garbage bag
- Flashlight or small desk lamp
- Pin
- Scissors
- Duct tape
- Old nylon stocking
- Embroidery hoop (optional)

Even through powerful telescopes the human eye can't focus on so distant a thing as a star. Stars exist so far from the Earth that when you look up, you don't see the star-objects themselves, but only the *refraction patterns* of their light across millions of miles. This project simulates the refraction patterns of distant stars by viewing pinpoints of light through a nylon-stocking gradient.

Procedure

1. Cut the poster board into the shape above (right column) and bend it around to make a cone, taping it as necessary. The narrow end of the cone should be wide enough to contain the flashlight.

2. Place the wide end of the cone on the floor, and tape the flashlight to the narrow end.

3. Place the entire cone in a black garbage bag, and pull the bag tight over the wide end of the cone like a drum. Twist the remaining bag down the length of the cone and tape it to the flashlight at a point just above the flashlight switch. Use the scissors to trim the excess bag away from the flashlight.

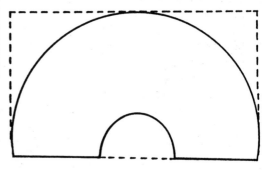

Poster Board Shape

4. Turn the cone on its side and make pinholes in the drum side of the cone. Avoid making too many holes close together, but space them evenly to recreate a starry night sky.

5. Cut the nylon stocking so that you have a piece slightly smaller than the area of the embroidery hoop. Stretch the nylon in the hoop.

6. Viewing starlight works best in a darkened area. Turn on the flashlight and stand about 10 feet (3 m) from the drum side of the cone.

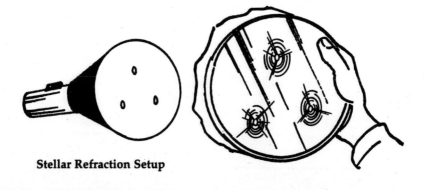

Stellar Refraction Setup

7. Hold the embroidery hoop and view the pinpoints through the stretched nylon. If you don't have a hoop then simply stretch the nylon in front of you.

Result: In the dark, tiny pinpoints of light appear on the wide end of the cone. Notice that these pinpoints already have a recognizable starlike shape as your eye struggles to focus. But when you look through the stretched nylon, a very clear refraction pattern emerges with pointed rays and pronounced halos surrounding each pinpoint of light—truly starlike!

Display Tip
Document the construction of your model with photographs and describe its operation. If possible, drape a dark piece of cloth that you can easily remove to reveal your model. In this way, the judges can both view the model and then experience the refraction patterns for themselves.

Did You Know?
When the light from stars reaches the Earth, it's already thousands, if not millions, of years old. The stars you see may actually no longer exist, and new stars may now blaze in the heavens that astronomers won't know about for a very long time.

Sheet Erosion

You Will Need

- Plastic planting tray with open bottom
- Sandy soil or garden soil mixed with sand
- 5 coins
- Watering can
- 5 metal lids from small cans

Soil Erosion Landscape

Sheet erosion means that water, running down from higher elevations, washes away the exposed particles of fine soil and leaves the protected coarser particles on higher plateaus. Sheet erosion formed the great sandstone buttes found in the southwestern parts of the United States. In the Painted Desert region of Arizona, for example, patches of rock-covered ground were protected from the eroding waters. After thousands of years, these rocky buttes now tower high above the surrounding, highly eroded landscape. This project recreates that landscape in miniature.

Procedure

1. Take the plastic planting tray outside and fill it to the top with the sandy soil.
2. Lay the lids and coins in various places over the soil, tapping them down a bit.
3. Fill the watering can, and gently sprinkle water over the tray. Allow the water to drain, and then resume sprinkling. Don't let the tray flood.
4. Allow the soil to dry, then repeat step #3.
5. Allow the soil to dry again. Carefully lift the lids and coins off the soil.

Result: The soil under the can lids and coins formed into little pedestals and plateaus.

Explanation

The protected particles of soil under the can lids and coins were shielded from erosion. As the soil around them washed away, the protected soil formed pedestals and plateaus.

Display Tip

Allow your sheet-erosion landscape to dry thoroughly before exhibiting it. If you're afraid that the delicate shapes in the soil will crumble, give the soil a coat of spray varnish, and allow the varnish to dry before transporting your landscape.

Did You Know?

When land developers expand beach areas by importing sand from other places, they must carefully match grain size. Different sizes of grains of sand make erosion worse, and a badly mixed sand combination will quickly wash out to sea. Matching grain size to existing sites is particularly important when constructing landfills.

Split a Stone

You Will Need
- Balloon
- Water
- Papier-mâché
- Brown and black paint
- Paintbrush
- Freezer

Is it possible for ice crystals to split stones and break mountains apart? Can even the tiniest trickle of water, when it freezes, crack open the mightiest of boulders? You can find out by constructing a "stone" from a water-filled balloon and papier-mâché. If you've never used papier-mâché before, have an adult help out with the first part of the project.

Part 1 Making a Papier-Mâché Stone

Procedure
1. Fill the balloon with water to about three-fourths of its size.
2. Mix flour and water into a thick paste, and add torn strips of newspaper.
3. Carefully layer each piece of paste-coated newspaper over the balloon, making sure that you completely cover the balloon in papier-mâché.
4. Allow your balloon to dry overnight or until the shell of papier-mâché becomes hard and brittle.

Part 2 Freezing the Stone

Procedure
1. To make your balloon more stone-like, paint the papier-mâché black and brown.
2. Place the stone in a freezer and leave it there for 24 hours.
3. Remove the stone from the freezer and examine its surface.

Result: The stone has been split open by the expanding ice in the balloon. In this way, rocks are split open by the expansion of ice in their crevasses.

Explanation
Most materials expand when heated and contract when cooled. But water expands when *cooled*, eventually forming ice crystals. Liquid water forms long strings of molecules that can easily seep into vacant places within rocks. When the seeping water freezes, the expanding ice crystals become a kind of wedge, and push the small spaces apart to break the rock.

Display Tip
Document the construction of your stone with photographs. Remember to take before-and-after photos of the stone. Exhibit your split stone for viewers to examine.

Did You Know?
Water seeping into crevasses and then freezing is the major cause of crumbling roads and pavements. Older surfaces are particularly affected, since normal wear and tear can create a cracked and uneven surface where water can accumulate and freeze.

The Parallax Principle

You Will Need

- Wooden plank 2½ feet × 6 inches (75 cm × 15 cm), ½ inch (1.25 cm) thick
- 2 protractors with center holes
- Ruler
- Pencil and paper
- Plastic straw
- Scissors
- Thumbtack
- Tape measure
- Epoxy glue
- Kite string
- 3 helpers
- Clear, flat area for measurements
- Masking tape
- Paper clip

The parallax principle for calculating distance was discovered by the Greek mathematician Aristarchus of Samos in about 270 B.C. From two observation points, the astronomer studies the object to be measured against background objects. What really interests him is the way the closer object seems to change position against the farther objects.

After measuring this change of position, astronomers use a method called *triangulation* to help them calculate distances. Triangulation means drawing an imaginary triangle that connects two observation points with the object to be measured. Astronomers can then use the known dimensions of the triangle to find the unknown dimensions—one of which will be the distance to the object.

Although simple in theory, calculating distance from parallax angles relies on complicated trigonometric formulas for accurate results. Instead, this project uses two triangulation tricks that give fairly accurate results while showing why parallax is so important to astronomers.

Part 1 Simple Triangulation Instrument

Procedure

1. Stand the wooden plank on its edge. Glue the ruler side of the protractor to the narrow edge of the plank. The protractor's circular side should stick out from the plank.

Protractor on Plank

2. Cut a 2-inch (5-cm) piece from the plastic straw, and then cut one side diagonally so that it tapers to a point.

3. Unbend a paper clip to make a long U-shape. Break off the extra parts of the paper clip.

4. Slide the paper clip on the straw about 1 inch (2.5 cm) from the tapered end so that the two prongs of the U make a sighter along the straw. Pinch the paper clip in place, or attach a narrow strip of tape to keep the sighter attached to the straw.

Paper Clip–Straw Sighter

5. Place the straw over the protractor so that the tapered end just barely overlaps the inside curve of the protractor. Make sure the two prongs of the sighter point up. Push a thumbtack into the straw and through the

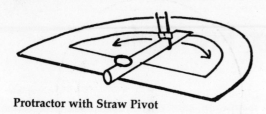

Protractor with Straw Pivot

center hole of the protractor. The straw should pivot freely against the protractor.

Part 2 Measuring Distance

Procedure

1. Find a clear, flat area. Have helper #1 be the Moon and stand quite a distance from you.

2. Use the measuring tape and masking tape to mark out a *line of position* opposite the Moon. The farther away your friend stands, the longer your line should be (to a maximum of 40 feet or 12 m). Make sure you measure this line of position accurately, and make the line longer or shorter to avoid fractions.

3. At the left end of the line of position mark an "A" with the masking tape. At the right end mark a "B."

Measuring Line of Position
Using Sighting Instrument with Two Helpers

4. Take your sighting instrument to the A end. Place the narrow edge of the plank directly over the line.

5. While helper #2 steadies the plank, crouch down and look along the straw, moving it as necessary, until you find the Moon between the two prongs of the sighter.

6. Have helper #2 record the degree reading from the bottom edge of the curved part of the protractor.

7. Take your sighting instrument to the B end of the line and repeat this procedure. But this time you'll take the degree reading from the top edge of the protractor.

8. With your measurements finished, go back to A and pick up the roll of kite string. While you hold one end of the string, have helper #2 walk to the Moon while unraveling the string.

9. When helper #2 reaches the Moon, pull the string tight so that you have a reasonably straight line between A and the Moon. At a signal from you, your helper can cut the string and roll it back up—your actual measurement of distance against which you'll eventually compare your calculations.

Part 3 Pencil & Paper

Procedure

1. On a piece of paper, draw your line of position. Scale it so that ¼ inch (.62 cm) equals 1 foot (30 cm), and mark ends A and B.

2. Move the protractor's center hole over A, and draw a line extending from A that corresponds to your earlier degree reading at position A.

3. Move the protractor over B and repeat the procedure. Where your lines meet indicate, to scale, where the Moon stood. Call this angle "C." *Note:* you can easily calculate the degree of this angle by adding up the angles at A and B and subtracting them from 180°. (A geometric rule states that all angles of a triangle must add up to 180°.)

4. Measure along side A–C of your triangle, and translate your measurement into feet. This represents the actual distance that separated you from the Moon.

5. To check your results, measure (in sections) the length of kite string against the tape measure.

Result: The kite-string measurement should approximate your conversion into feet (or centimeters, if you are using metrics). Although this method is crude compared to trigonometric calculations, it gives you an idea of how the parallax principle works.

Explanation

To understand the general formula behind the parallax principle, look at the curved part of your protractor again. Remember, the circle is just as important as the triangle in calculating distance. In the diagram you can see that your triangle grows from the center of an imaginary circle. Better yet, imagine this triangle moving *around* the circle like the minute hand of a clock. Remember, a circle has 360°.

Call the dark area between the two long sides of the triangle "A," for *angular diameter*. Call the short side of the triangle "L," for *line of position*. As you imagine the triangle moving about the circle, notice how A sweeps through the *circle's area* while L moves along the *circle's circumference*. This means that A is

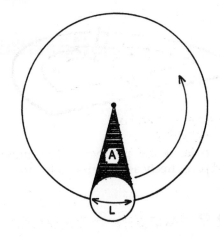

Area of circle = 360°
Circumference of circle = pi (3.14)
Angular diameter = A
Line of position = L

The *angular diameter* (**A**) is the same fractional amount of 360° (area) as the *line of position* (**L**) is of the circumference.

related to the circle's area in the same way that L is related to the circle's circumference. Or, to put it another way:

*A and L are equal fractions
of the whole circle.*

This is why accurate degree measurements are so important in figuring out the dimensions of triangles.

Parallax Shift

You Will Need

- Masking tape
- Tape measure
- 3 helpers
- Binoculars (optional)
- Clear, flat area for taking measurements

You can combine this project with The Parallax Principle.

Surveyors and astronomers alike use the parallax shift to measure distances to distant objects. From two observation points, the person views the closer object against background objects. What really interests him is the way the closer object seems to change position against objects farther away. From this degree of shift, he can then calculate the approximate distance of the object.

This project uses nothing more than masking tape, a tape measure, and a few friends to demonstrate the ingenuity of this ancient method of measurement.

Procedure

1. Tape a line of position at least 7 feet (2 m) long with A and B at the ends and X in the center.
2. Have a friend, the Moon, stand at some distance opposite the X. Don't measure this distance.
3. Well behind your Moon, have two helpers stand about 50 feet (15 m) apart. These helpers will be star #1 and star #2.
4. Stand on the X and observe your Moon against the background stars (fig. 1).
5. Move to the B end of the line of position, and observe the Moon's apparent shift against the stars (fig. 2).
6. From B, direct star #1 to walk to the right until completely hidden by the Moon.

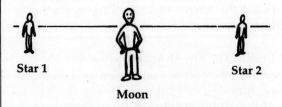

Figure 1. View from X (step 4)

Note: A shared pair of binoculars will help your stars see your directions more clearly.

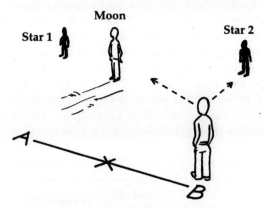

Figure 2. View from B (step 5)

7. Move to the A end of the line of position, and again observe the Moon's shift (fig. 3).

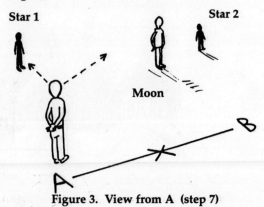

Figure 3. View from A (step 7)

8. Now direct star #2 to walk to the *left* until completely hidden by the Moon.

Parallax Shift

9. Have your stars measure this new distance between them with the tape measure.

10. Calculate the distance with the Distance-to-Moon formula shown below. The line of position (L of P) divided by the new distance between stars (N D) equals the distance to the moon.

$$\frac{L\ of\ P}{N\ D} = \text{Distance to Moon}$$

If the line of position is 7 feet long) and the new distance between stars is 42 feet, then 42 divided by 7 equals 6 feet—the distance from X to the Moon. Study the diagram below.

In metrics, let's say the line of position is 2 meters long and the new distance between stars is 15 meters, then 15 divided by 2 equals 7.5 meters—the distance from X to the Moon.

Display Tip

You'll probably have fun using these parallax tricks with your friends or classmates. Make sure someone takes photos of the activity so that you can display them at the exhibit. Draw out your calculations clearly on a large piece of poster board so that all your observers can understand the parallax principle as well as you.

Did You Know?

Astronomers also use the parallax principle to measure distances to stars. But stars are such distant objects that the line of position must be much greater than the diameter of Earth. To solve this problem, astronomers use the diameter of the Earth's orbit around the Sun. This gives them a line of position long enough for calculating greater distances.

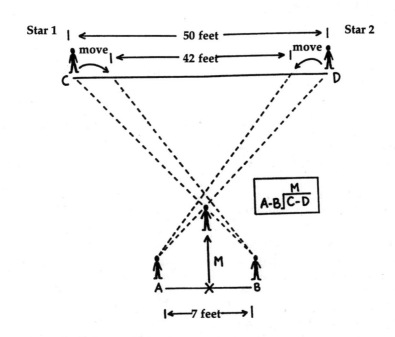

Measuring Distance to the Moon from View X
Using Star Positions and Formula

Height Calculation

You Will Need
• Scientific (not business) calculator
• Ruler
• Protractor
• Pencil and paper

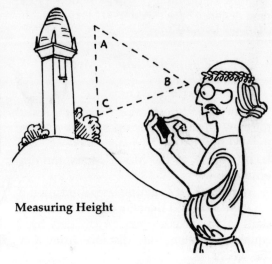

Measuring Height

The work *trigonometry* comes from the Greek words *trigōnon* and *metria* meaning "measuring triangles." Long ago, the ancient Greeks discovered that it was useful—if not plain fun—to figure out the relationships between the sides and angles of a right triangle (which is a triangle with one 90° angle), and then to express these relationships in a formula..

The Greeks understood that, no matter what the size and shape of a right triangle, certain relationships always exist. It was therefore possible to figure out the unknown length of one side of a right triangle if you knew the lengths and angles of the other sides.

Of course, the Greeks had to manually compute their formulae in order to come up with the correct answers. Today, the handy calculator speeds things along, and you can easily compute the sides of a right triangle by using your calculator's tangent function, or T.

Procedure

1. Using the ruler and protractor, draw a right triangle like the one above right, labeling the corners A, B, and C, and labeling the sides "opposite," "adjacent," and "hypotenuse." Use your protractor to make sure that the angle at corner B is 35°. Because this is a right triangle, you know already that the angle at corner C is 90°.

2. Imagine that this is a large triangle, so side B–C is 100 yards long, and that side A–C represents the unknown height of some tall tower.

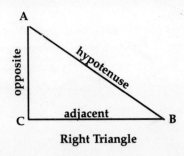

Right Triangle

3. Find the height of the tower (line A–C) by entering this formula into your calculator:

Enter: 100
Press: or * (times)
Enter: 35
Press: Tan (for tangent function)
Press: =

4. The result should be 70.0, for 70 yards. This means that, given an angle of 35°, the opposite side to this angle is always 0.7 times the length of the adjacent side.

Did You Know?

You can use the tangent function on your calculator to find other sides of a right triangle. To find the adjacent side of the triangle when you know the opposite side is, say, 70 yards, divide the length of the opposite side by the tangent of 35° Here is the operation on your calculator:

Enter: 70
Press: (division sign) or /
Enter: 35
Press: Tan
Press: =

The result is 99.97, which you can round off to 100 yards.

In metrics, if, say, the opposite side is 70 m (that's 77 yards); divide the length of the opposite side by the tangent of 35°. And the result would be about 100 meters.

Rare-Earth Metals

You Will Need

- Horseshoe (U-shape) magnet
- String
- Plastic sandwich bag
- Twist fastener
- 2 wide-mouth glass jars
- Garden spade
- Paper towel
- Magnifying glass
- Soil
- Wooden paint stirrer

Dipping Magnet into Jar #1 with Water and Soil (step 4)

The soil from your garden hides minute particles of magnetic material. An easy way to find and study some of this stuff involves using a horseshoe magnet to sift through samples of dirt. This method can reveal traces of *rare-earth metals*, most of which lie deep underground.

Procedure

1. Cut a 12-inch (30-cm) length of string and tie one end around the U-shape magnet. Make sure that the ends, or poles, of the magnet point straight down.
2. Place the plastic sandwich bag over the magnet, pulling the bag tight across the magnet's poles. Use the twist fastener to close the bag and keep the magnet watertight.
3. Fill both jars about three-quarters full of clean water. Place 3 spades full of soil in the first jar, stirring the soil and water mixture with the paint stirrer.
4. As the soil swirls through the water, dip the bagged magnet into the jar, dropping it to the bottom and lifting it to the top again with the string. Do this several times or until the soil begins to settle in the jar.
5. Carefully lift the magnet from the jar and look at the ends.
6. Dip the magnet into the second jar of clear water, and carefully remove the plastic bag from the magnet so that the particles drop to the bottom of the jar.
7. Reattach the plastic bag, and repeat this operation until your magnet no longer attracts particles from the soil and water mixture.
8. Carefully pour off the water from the jar containing the particles until you have just a little water remaining at the bottom.
9. Pour this remaining water and particle mixture over a paper towel and let the towel dry. Then observe the dry particles through the magnifying glass.

Result: The first time you dip the

Dipping Magnet into Jar #2 with Water (step 6)

magnet into the jar, small gray, black, white, and red particles stick to the plastic bag, attracted to the magnet. When you dip the magnet into the second jar of clean water and remove the plastic bag, the particles drop off. Repeating this process several times removes most of the magnetic particles in your soil sample. Allowing the particles to dry on the paper towel makes it easier to see them.

Explanation

The particles belong to a class of magnetic materials called *ferrites*. Your collection probably includes magnetite (natural iron or lodestone), manganese, magnesium, zinc, barium, strontium, ferric oxide, and tiny pieces of such strange-sounding materials as hematite (an iron ore), ilmenite (a titanium ore), and yttrium-iron garnet (red particles).

Not only are all these metals formed in igneous (made by heat) rocks, but since you found samples in your own backyard, you can be pretty sure that larger metal deposits exist close by.

Display Tip

Exhibit your samples of rare-earth materials on a clean piece of blotting paper with a magnifying glass close by.

Did You Know?

By studying the Earth's magnetic field, scientists can collect important information about the composition of soil. For example, geologists can find valuable mineral deposits underground by placing *magnetometers* aboard airplanes. These sensitive instruments point out odd "blips" in the Earth's magnetic field that may indicate a large mineral deposit.

Ordinary soil contains rare-earth metals.

Magnetic Filtration

You Will Need

- Rusty piece of iron
- Fine sand
- Fine sandpaper
- Sheet of white paper
- Old teaspoon
- Saucer
- Wooden stirrer
- Horseshoe magnet
- Small plastic bags
- Twister ties
- Small bowl of water

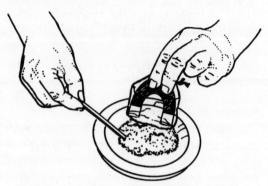

Dipping Magnet into Rust-and-Sand Mixture (step 5)

Geologists often have to separate materials in the earth that may look the same but have different properties. This is particularly true for *ferromagnetic* materials, or materials that have magnetic properties. Ferromagnetic particles look the same as sand or clay with one important difference: they contain a natural form of iron called *magnetite*.

This project shows how to make a simple magnetic filter to separate iron particles from sand.

Procedure

1. Place the white sheet of paper on a flat surface. Use the fine sandpaper to scrape rust from the iron onto the paper until you have a about 1 teaspoonful (5 ml).
2. Fold the paper and pour the powdered rust into the saucer.
3. Add 2 tablespoons (30 ml) of sand, and combine the mixture with the wooden stirrer.
4. Place the plastic bag over the end of the horseshoe magnet, and tie the bag closed with a twist tie.
5. Dip the magnet into the rust-and-sand mixture, and gently push the mixture around with the stirrer. Particles of iron and iron rust will stick to the plastic bag.
6. Remove the magnet from the mixture, and carefully dip the bag-covered end of it into the bowl of water.
7. While still holding the bag in the water, untwist the tie and remove the magnet from the bag.
8. Gently swirl the bag around in the water until all the particles have dropped off.
9. Repeat steps 4–6 with fresh plastic bags until you remove all the iron particles from the sand.

Result: The particles of iron, attracted to the magnet through the plastic bag, separate from the nonmagnetic sand particles. Dipping the magnet into the bowl and then removing it from the plastic bag allows the iron particles to drop from the bag and sink to the bottom of the bowl. If you boil the water in the bowl until it evaporates, you can collect the iron particles again and display them next to the sand.

Did You Know?

People who work in food-processing factories know the danger of metal fragments winding up in our food. Since farmers have the same concern for their livestock, they invented something called a *cow magnet*. This is a small, rounded magnet the cow swallows when just a calf. The magnet is very small and very smooth; that way it doesn't hurt the calf at all and may even save its life. The magnet attracts small pieces of wire the cow might swallow while grazing and keeps the pieces from puncturing the cow's stomach.

Box Periscope

You Will Need

- Poster board 11 × 14 inches (27.5 × 35 cm)
- 2 pocket mirrors
- Craft knife
- Metal ruler
- Scissors
- Tape
- Jar lid

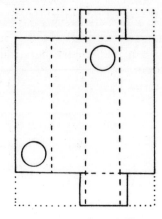

Pattern for Box Periscope

Mirrors, important to astronomers, can also prove useful for more earthly applications. The reflecting properties of mirrors allow us to extend our sense of sight, and complicated arrangements of mirrors can allow us to see around, over, and even behind objects. The submarine operator uses a variation of this simple box periscope to see above water. And physicians use an advanced form of the periscope, called a *fiber-optic tube*, to look inside the human body.

This project shows how to make a box periscope from pocket mirrors and poster board. It'll come in handy the next time you have to peer over a high fence or around a corner.

Procedure

1. Place the poster board on a flat surface and copy the patterns according to the diagram (above right), including dotted lines. Use the jar lid to trace the circles at each end of the pattern.

2. Cut around the solid lines only, including the circles. Don't cut around the dotted lines; they indicate where to fold the poster board.

3. Ask an adult to help you for this part. Place the metal ruler against each dotted line, and make a shallow slice along the line with the craft knife. This will help you fold along the dotted lines.

4. Turn the pattern over and carefully fold it.

5. Close the box along the folds, leaving the top and bottom flaps open. Stretch a long piece of tape where the edges join. Place the box on its back with the top hole on your right side.

6. Attach a strip of tape to the edge of one of the pocket mirrors. Carefully slide the mirror into the box's end with its reflecting side up, until you see only the mirror when looking through the hole.

7. Press the tape down to anchor the mirror in place.

8. Turn the box end up, with the mirror at the top. Gently push the mirror so that it falls forward at a 45° angle toward the hole.

9. Repeat steps 6-8 for the second mirror on the opposite side of the box.

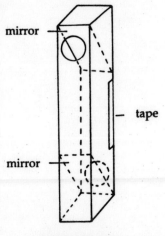

Box Periscope
(*inside view*)

Result: Your box periscope represents a basic periscope design: a hollow shaft with reflecting mirrors placed opposite each other. The light, entering one side, makes two right angles before it reaches your eyes on the other side. A second advantage of this arrangement is that the second mirror corrects the reversed image of the first mirror, so that you can easily read signs through your periscope.

Display Tip
Make two periscopes for your display. One should be for demonstration purposes and sturdy enough to allow judges and viewers to use it. The second periscope should be your model. A cutaway diagram of your periscope will help viewers understand how a periscope works.

Mapping the Ocean Floor

You Will Need
• Small table
• Two chairs
• Desk bell (bell with button)
• Graph paper
• Pen
• Watch with a second hand
• Friend to help

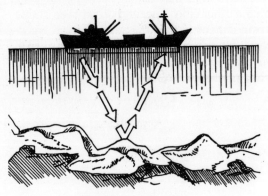

Mapping Ocean Floor

Scientists use sound to study the shapes of things they cannot directly observe. When sending a sound toward a distant object, scientists know that the sound will bounce off the object and eventually return to them. This technique, called *echo-location*, is similar to what a bat does naturally to find its way around in the dark. But do you think that echo-location could help scientists study a very large object like the ocean floor? This project will help you find out.

Part 1 Listening for Pings

Procedure

1. Sit opposite your friend at the small table. Pretend this is the navigation room of a ship sent to map an unknown portion of the ocean floor.
2. Decide who will ring the bell and who will record results. The bell ringer represents the echo-location device.
3. The ship sends ping #1 to the ocean floor. Using the watch, you must time in seconds (and write down) how long it takes the ping to echo back. For now, the second ping of the bell represents the echo.
4. The ship sends out ping #2, and now you must time the interval between this ping and its echo.
5. Proceed for eight more pairs of pings, each time recording the time interval between ping and echo.

Part 2 The Echo Equation

Procedure

1. Use each of your time intervals in the equation below. The equation represents the time it takes for the ping to reach the ocean floor multiplied by the speed of sound in water (1,500 meters per second).

Time (divided by 2) × Speed of Sound in Water = Ocean Depth

For example, if 4 seconds elapsed between ping #1 and its echo, then it took 2 seconds for the ping to reach the ocean floor.

2 seconds × 1,500 meters per second = 3,000 meters

2. Continue to translate each of your time intervals into meters until you reach the eighth and last ping.

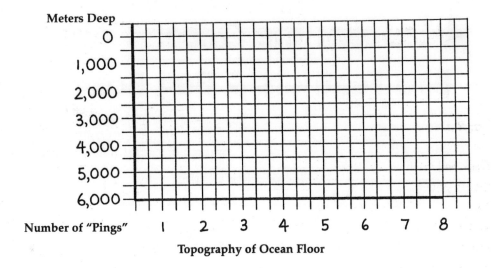

Topography of Ocean Floor

Part 3 From Figures to Pictures

Procedure
1. Copy the chart on p. 41 on a piece of graph paper. Draw a dot at the meter depth that corresponds to each of your pings. Remember, each ping is actually a pair of pings, one representing the echo.
2. Connect the dots on the graph paper.

Result: A line with peaks and valleys emerges. This line represents the transformation of your time-interval data into meters-deep data.

Explanation
By connecting the dots, you've transformed numerical data into a graphic depiction of an imaginary ocean floor. The topography of this ocean floor was determined entirely by the intervals between each ping and its echo. Your friend, the bell ringer, can create any landscape he chooses by varying these interval times.

Display Tip
Make many ocean floor explorations and display them all. Clearly exhibit your equation for determining meters from time interval measurement. Research and display any interesting information you find about echo-sounding devices on ships.

Did You Know?
Although some parts of the ocean floor are flat, like the Great Plains of the United States, other parts of the ocean are dotted with mountains and valleys. Most of the mountains are volcanoes, some so tall that they rise above the ocean's surface to form islands as they erupt. The valleys, also called trenches, are narrow and deep.

Although trenches are found in every ocean, the deepest trenches scientists know about are located in the Pacific Ocean. The Marianas Trench is nearly 11,000 meters (about 36,000 feet) deep—that's 2,000 meters (about 6,500 feet) deeper than Mount Everest is high.

Bird Feathers & Bug Boxes

Ant Architecture
Berlese Funnel for Soil Parasites
Floating Feather
Trap for Nocturnal Insects
Hummingbird Feeder
Marching Trail of Ants
How Is a Butterfly Like an Elephant?
Be a Web Master
Bird Pudding
A Bug's Favorite Color

Ant Architecture

You Will Need

- 2 wide-mouth glass jars
- Frozen-orange-juice can or similar can
- Small dish
- Fine cloth netting
- Rubber band
- Digging spade
- White paper
- Soil sample with ants
- Gardening gloves
- Small block of wood
- Pie tin
- Black construction paper
- Cellophane tape

This project lets you observe the elaborate tunneling structures of an ant colony and determine the social organization of ants. All you'll need are two glass jars, a frozen-orange-juice can, a pie tin, soil, and, of course, ants.

Procedure

1. Collect worker ants by using the digging spade to gently lift soil under flat rocks. Place the soil sample on the white paper and gently stir it. As the ants scatter, fold the paper and brush both soil and ants into the jar. Replace the jar lid without screwing it on.

   Caution: Use gardening gloves when collecting ants. Some species bite.

2. Continue digging until you see ants scattering with larvae. Take one final clump of soil and place it on the white paper. As you gently break it up, you should see the much larger and paler ant queen emerge. If she doesn't appear, take another soil sample and repeat the procedure.

3. Deposit the queen in the jar and screw the lid on. Take the jar to where you'll set up your colony.

4. Place the frozen-orange-juice can in the center of jar #2. Remove the lid from jar #1 and use a tablespoon to transfer the soil and ants to jar #2. Be extra careful that you don't harm any ants, particularly the queen, during this procedure.

5. When you've surrounded the orange-juice can with ants and soil, rest a small water-filled dish on top of the can. Place netting over the mouth of the jar and secure it with the rubber band.

6. Partly fill the pie tin with water, and place the block of wood in the center.

7. Rest the ant-colony jar on the block of wood so that the water in the pie tin forms a kind of moat around the jar. This will keep ants clever or small enough to squeeze through the netting from escaping.

8. Make a wide tube of black construction paper to fit snugly over the jar. Place the tube over the jar, and leave your ants undisturbed in a warm location for 24 hours.

9. Remove the tube to watch the ants; add bits of bread occasionally to feed your colony.

Result: The ants construct an elaborate connecting network of tunnels close to the inner surface of the jar. Some tunnels appear to end in small chambers where larvae are kept.

Explanation

Calmed by the presence of their queen and no longer feeling threatened, your ants waste no time in setting up house. Ant colonies have an elaborate structure of tunnels, chambers, nurseries for larvae, and even "gardens" of nutritious molds. As your colony develops, so will the complexity of the tunneling system. You may even notice an ant "cemetery" after a while! The social order of ant colonies usually include three classes: winged, fertile females; wingless infertile females, or workers; and winged males. In some species, workers may become soldiers or other specialized types.

Without the black construction paper, the ants would tunnel toward the jar's interior rather than close to its surface.

Display Tip

Document your ant collecting with photographs. Display the living colony in your booth. Use Post-It notes attached to your jar to identify some of the colony structures, such as tunnels, chambers, or nurseries. If you're lucky enough to have the queen exposed, clearly identify her.

Did You Know?

Not all ants live in tunnels. Some species of ants live in mounds they build above the ground, and other species live in wood. Army ants are mostly on the move, traveling in columns and destroying plants and animals in their way. When they do stop briefly, they live in tangled ant structures made of their own bodies.

Berlese Funnel for Soil Parasites

You Will Need

- Poster board
- Piece of cloth netting (not too fine)
- Masking tape
- Large glass jar
- Liquid dishwashing detergent
- Tablespoon
- Directional lamp
- Digging spade
- Pail
- Soil sample

Soil-burrowing insects can destroy a garden overnight. Hiding under rotted leaves and other debris, or lying just under a thin layer of soil, these insects will emerge in the dark to consume an entire tomato plant or flower garden. Unless you sit up all night with a flashlight, you'll find it very difficult to spot the bugs that do this nasty munching. But with the aid of a cleverly designed Berlese funnel, named for its inventor, you can find and trap them.

Procedure

1. Cut the shape below from the poster board, roll it into a wide cone, and tape the ends together.

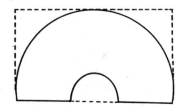

2. Cut the cloth netting to just fit over the narrow part of the cone, and tape the netting to the sides of the cone.

3. Fill the large jar with water, and add about 1 tablespoon (15 ml) of dishwashing detergent.

4. Place the narrow end of the cone in the mouth of the jar so that the cone sits upright, supported by the jar.

5. Use the digging spade to fill the pail with soil samples. Take samples from under shrubs or large, shady clusters of plants.

6. Empty the soil into the cone.

7. Place a strong directional lamp over the wide end of the cone, close to the soil.

8. Leave the light on all night.

Result: In the morning, you'll find that many insects have dropped into the water-filled jar. These insects represent a good sampling of the type of soil-burrowing pests that exist in your garden.

Explanation

Since soil insects hate the light and heat, they burrow down through the soil and fall through the cloth netting.

Display Tip

In addition to your funnel, display photographs of your garden and where you took the soil sample from. Document the damage to your garden. Buy an insect guide for your area (or check one out from the library) and try to identify some of insects you've captured. Earwigs, cut worms, and weevils are all familiar soil parasites.

Floating Feather

You Will Need

- 2 store-bought white feathers *(Do not use a found feather.)*
- Wide bowl of water
- 2 teaspoons (30 ml) of liquid dishwashing detergent
- Blue food coloring
- 2 cotton swabs
- Tweezers
- Paper towel

This project demonstrates the effect of detergent pollution on the buoyancy of freshwater birds. With the overuse of soapy detergents, soap-contaminated water eventually finds its way into natural groundwater reservoirs. Natural bodies of water, such as ponds and lakes, are replenished by this polluted groundwater.

Procedure

1. Fill the bowl with water and add blue food coloring.
2. With the tweezers, gently float the first feather on the water.
3. After 1 minute, gently run the cotton swab across the surface of the feather. Examine the swab.
4. Remove the feather and place it on a paper towel.
5. Add 2 teaspoons (30 ml) of dishwashing liquid to the water. Stir the water gently to avoid making bubbles
6. Carefully float the second feather.
7. Wait 1 minute, and run a clean cotton swab across the surface of the feather. Examine the swab.
8. Remove the feather, and place it on a paper towel to dry.

Result: Both feathers float, but the feather placed in the soapy water allows water to come through and soak the dry side. You could see this when you stroked that feather with the swab, and the tip of the swab picked up the blue color of the water.

Explanation

To understand what happened, we have to look more closely at feathers. The "stem" of the feather and the part that attaches to the bird's body is called the *quill*. It's mostly hollow and so helps the bird float. Branching out from the quill are the fine feathery barbs that have hooklike barbicels along their sides. These barbicels lock together to form a continuous surface.

Adding soap to the water breaks the surface tension so that water can flow through the tightly locked barbicels. Soap also dissolves the natural oil on the feathers so that the bird becomes "heavy" in the water, less able to propel itself, and more likely to tire out. Ducks, swans, and egrets, as well as freshwater otters and beavers, can all drown in soap-polluted water.

Display Tip

Document your procedure with drawings and photographs. Display this experiment so that others can participate and share your conclusions.

Did You Know?

A light coating of sticky oil is the reason a natural feather duster works. The same holds true for natural wool dust rags. But in the case of wool, the oil, *lanolin*, is also used as a base for many cosmetics.

Trap for Nocturnal Insects

You Will Need

- Medium-size cardboard carton with flaps
- Aluminum foil
- Bright flashlight or battery lantern
- Strong cord
- Old sock
- Rubber band
- Duct tape
- Glass jar
- Brass fasteners
- Scissors
- White paint
- Paintbrush
- Ruler

This trap for nocturnal insects will allow you to capture insects without the use of harmful chemicals. The trap will provide a good sampling of the kind of insects that exist in your area, and the results will vary according to the season, temperature, and location or site.

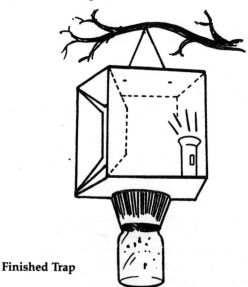

Finished Trap

Procedure

1. Paint the outside and inside of the box white, except for the inside back wall. Allow the paint to dry.
2. Coat the inside back wall of the box with rubber cement, and line it with a sheet of aluminum foil.
3. Trim each of the four flaps, following the diagram below.

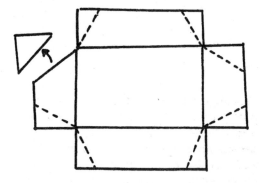

Trim box flaps. (step 3)

4. Push the flaps inside the box so that they come together in a kind of picture-frame design with the cut edges joining. Don't tape the flaps together because you'll need to push them aside to construct the rest of the trap and insert the flashlight.
5. With the box turned on its side so that the flaps face you, cut a hole about 4 inches (10 cm) in diameter in the "floor" of the box.
6. Cut off the toe end of an old sock, and stretch one end of the sock in the hole. Secure the sock to the edges of the hole by pushing brass fasteners through both sock and cardboard. You should wind up with a kind of sock tube that leads out of the box.
7. Attach the other end of the sock to the lip of the glass jar, and secure it there with a rubber band, twisted several times.
8. Make two holes (as small as possible) in the roof of the box, centering the holes toward the back wall of the box. The position of the holes has to counteract the weight

Trap for Nocturnal Insects

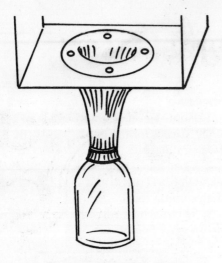

Attach sock to lip of jar. (step 7)

of the flashlight to ensure that the box hangs level. String the cord through the holes.

9. Insert the flashlight or lantern so that as much light as possible reflects off the aluminum foil. Close the flaps of the box to make a picture frame again, and hang your box in a tree so that the sock and glass jar hang beneath. It's a good idea to have your trap as far away from any other lights as possible.

10. Check the jar periodically. If necessary, leave the trap out all night and observe your specimens in the morning.

Result: The jar will fill up with nocturnal insects—particularly moths and mosquitoes.

Explanation

Since insects use the Sun and Moon to navigate, they will move toward any bright light source. Attracted to the light inside the box, insects enter the box but cannot exit easily due to the flap design. Many insects, like moths, like to keep the light to one side of them, and so, they fly in circles around a light until they become exhausted. After fluttering for some time, your bugs drop through the sock into the bottle.

Display Tip

Take photographs to document the construction of your trap, and exhibit the finished trap in your booth. Make sure to take a photo of the trap in its environment, including such information as time of night, weather conditions, and how long the trap remained hanging.

Research the insects you capture and exhibit live insects, if possible. Many insect species, such as the praying mantis, are very beneficial and should not be harmed. Insect guides are an excellent resource for identifying insect life in various geographical locations. You can easily display insects in a screened jar, placing a small piece of sugar-soaked bread for food.

To expand your project, take your insect trap with you to various areas and compare the different sorts of insects you collect. You'll find that nocturnal insect life varies from area to area, particularly when there's water close by.

Did You Know?

Insects see the world very differently than mammals. Many insects can only see high and ultra-high spectrum colors such and blue, violet, and ultraviolet. This kind of color perception allows them to see the Sun through cloud cover and helps them navigate. Scientists also suspect that some insects may have amazing pattern-recognition abilities in order to recognize various flowering plants, even at great distances.

Hummingbird Feeder

You Will Need

- Clear plastic dishwashing liquid detergent bottle
- 2 plastic straws
- Large nail
- Sharp pencil
- Epoxy glue (available at hardware stores)
- Sugar water
- Red food coloring
- String

With a clear dishwashing-liquid bottle and a few plastic straws, you can easily make a hummingbird feeder and demonstrate the effects of atmospheric pressure at the same time. The clear bottle works best because the birds will be attracted to the red color of the sugar water as well as to the water's scent.

Procedure

1. Wash out the dishwashing-detergent bottle thoroughly to remove all traces of detergent. Keep the squeeze cap—you'll use this later.
2. Have an adult help you punch four equally-spaced holes near the bottom of the bottle, using the nail. If your bottle is squarish, punch two holes on each opposite side.
3. Enlarge the holes with a sharp pencil, but insert the pencil at a steep angle (toward the bottle's bottom) so that the hole becomes somewhat oval-shaped. This shape is necessary for inserting the straws at an angle.
4. Cut the two straws in half, and carefully insert all four pieces into the holes you just made. Push each straw two-thirds into the bottle, making sure that each straw angles up.
5. To keep the hummingbird feeder from leaking, apply epoxy glue where the straws enter the bottle. Allow the glue to dry overnight.
6. For the solution, mix 1 cup (240 ml) of sugar in 2 cups (480 ml) of water, and add a few drops of red food coloring.
7. Before you begin filling your feeder, place it over the sink. Insert a funnel into the mouth of the bottle.
8. Quickly, but carefully, pour the solution into the feeder through the funnel. The solution will begin to spill out through the open straws until you replace the closed squeeze cap.
9. Secure the cap tightly and tie string around it. Hang your finished hummingbird feeder close (but not too close!) to a window where you can enjoy the show.

Result: The solution travels to the tips of the straws but does not spill out. As the hummingbirds feed, more solution travels from the bottle to the straws without overflowing.

Explanation

A hummingbird feeder relies on atmospheric pressure to keep most of the liquid inside the bottle and just enough at the tips of the feeder pipes to attract hummingbirds. Before you replaced the cap on the bottle, more atmospheric pressure acted upon the liquid in the bottle than on the tips of the straws, and so the solution overflowed. But by securing the cap on the bottle, pressure inside the bottle was lessened so that the greater pressure came from outside.

Display Tip
Document the construction of your feeder with photographs. Try to photograph hummingbirds feeding. Display your model next to an explanation of why it works.

Marching Trail of Ants

You Will Need

- Shallow cardboard box with cover
- Glass or Plexiglas from picture frame (must completely cover box)
- White paint and paintbrush
- Cellophane tape
- Scissors
- Garden trowel
- Jar with cover
- Garden gloves
- Saucer
- Banana
- Granulated sugar

Have you ever had a trail of ants practically carry off your sandwich at a picnic? How did they find the sandwich to begin with? This project gives you a few clues.

Procedure

1. Remove the cover of the box, and paint the inside of the box white.
2. Cut the long side from the box cover, and make two notches, about 6 inches (15 cm) apart, along one edge.
3. Fold the ends of this piece so that it fits snugly into the box and divides the box in half. Tape it in place.
4. Look outdoors for an active anthill. Put on your garden gloves and use the trowel to make one deep cut into the ground, about 2 inches (5 cm) from the anthill opening. Quickly spill the soil into the jar and screw on the cover.
5. Cut a small piece of banana and place it a saucer. Sprinkle some granulated sugar over the banana, and then sprinkle some water over the sugar. Place the saucer in half of the box, near a corner.
6. Remove the cover of your jar, and quickly spill the soil ants into the other half of the box. Immediately place the glass cover over the box.

Result: At first, ants will race around wildly. But after about 20 minutes, they'll calm down and begin to explore their new environment. Soon a few ants will find the notches in the divider and cross to the half of the box containing the banana. In about an hour, you find a long trail of ants marching to and from the banana snack through the notches.

Explanation

Ants have a complex scent-based communication system that uses chemicals called *pheromones* that are produced in their bodies. Usually, one ant discovers a food source, the sugar-coated banana in this case, and leaves a pheromone trail to guide other ants to the

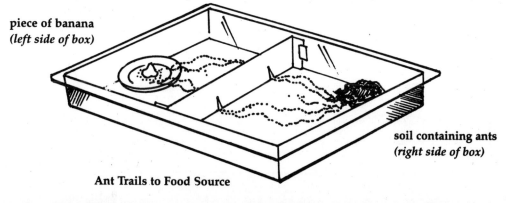

piece of banana
(left side of box)

soil containing ants
(right side of box)

Ant Trails to Food Source

food. As more and more ants walk the pheromone trail, they add to the scent. Interesting enough, if the original ant strays, or meanders over a pebble or twig, the other ants dutifully follow, even when a more direct route would seem more efficient.

Finding food is only one pheromone signal. Pheromones can also function as complicated chemical messages. If you bother an ant, alarming it, the ant immediately sends out a pheromone signal telling ants within a certain radius to flee from danger. But beyond this radius, the signal's chemistry changes to tell warrior ants to move in and attack.

Display Tip
Show off your ant box and pheromone trail, replenishing the food as necessary. The ants will survive about a week without their queen but will die shortly after that.

Did You Know?
Ants are only one species of insect that use chemical signals to communicate. The moth, for instance, uses feathery antennae called *chemoreceptors* to detect the scent of a moth of the opposite sex. These receptors are amazingly powerful and can detect the scent of another moth for miles.

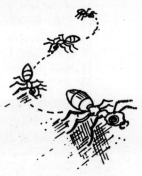

How Is a Butterfly Like an Elephant?

You Will Need

- Small glass aquarium
- Screen or piece of cotton cloth to cover top of aquarium
- Windowpane magnifying gradient (available at hardware stores)
- Butterfly net
- Glass jar with lid
- Modeling clay
- Wooden dowel
- Ripe banana
- Granulated sugar

Of course, a butterfly doesn't resemble an elephant, but watching a butterfly feed might remind you of one. Insects have specialized structures for mouths, perfectly suited for the kinds of foods they eat. For example, both the ant and grasshopper have powerful *mandibles* that slice through tough plant tissue. The nectar-feeding butterfly has the strangest mouth of all, a long tube called a *proboscis* that it keeps tightly coiled until it is hungry. When a butterfly senses something sweet, it uncoils its proboscis and pokes it around, looking for the source of sweetness. When found, the tube then sucks up the butterfly's lunch.

The elephant's long flexible snout, which we commonly call the trunk, is also known as a proboscis.

Procedure

1. If you don't have a butterfly net, you can make one by sewing an old pillowcase around a piece of wire, bent into a circle. Tape the wire around the end of a 1-foot (30-cm) wooden pole.

2. Capture a few butterflies and place them in your glass jar. You can remove the butterflies from the net without harming them by gently pinching their wings together.

3. Prepare your aquarium by placing a chunk of clay inside at the bottom, near the glass. Push the wooden dowel into the chunk of clay so that the dowel stands straight.

4. Mold a small cup from the remaining clay, and stick the clay cup at the top of the dowel.

5. Place a piece of ripe banana in the cup, sprinkle it with a little granulated sugar, then sprinkle the sugar with a little water.

6. Place the magnifying gradient outside the aquarium, on the glass nearest the dowel.

7. Release your butterflies into the aquarium, and cover the aquarium with the screen or sheet.

Result: At first, the butterflies will flutter around wildly. But after about a minute, they'll calm down and begin to explore their new environment. Eventually, one of the butterflies will find the sugar-coated banana and land at the edge of the cup. Slowly move your face to the side of the aquarium and look through the magnifying gradient. It might take a few tries and some patience, but sooner or later you'll see your butterfly uncoil its proboscis and begin to snack.

Explanation
The butterfly senses a soft, sweet food like nectar—its favorite dish. It uses its proboscis first to investigate the food source and then to suck it up. If your banana is soft enough, notice how it begins to disappear as the tube moves from place to place.

Display Tip
Show off your aquarium and feeding butterflies. You shouldn't plan to keep the butterflies in captivity for more than a week. Release the butterflies at the end of your project display period.

Be a Web Master

You Will Need

- Sheets of black construction paper or black poster board for larger webs
- Scissors
- Garden gloves
- Spray glue
- Spray varnish
- Willing assistant

This project will make you a web master—or at least a master at collecting spiderwebs. The best season for spiderwebs is early spring, and the best time to find them is early in the morning.

Before you go web-collecting, you should know that some spider species are dangerous. If you see an orb web, you should look along the edge for a spider. Find an insect field guide for your area, and try to identify the spider before disturbing the web. Many poisonous-spider species build orb webs, including the black widow. Triangle webs and cobwebs usually belong to the much safer garden-spider variety of the order Araneae.

Procedure

1. When you find a spiderweb specimen you like, put on your garden gloves, and carefully hold a piece of black construction paper behind the web and against it. Have a friend hold the spray glue and scissors for you.

2. Spray a thin film of glue over the web and construction paper, then quickly cut the threads around the outside of the construction paper to release the web.

3. Allow the glue to dry, then give your preserved web a light coating of spray varnish. Let the varnish dry overnight.

Result: You will have a clear, perfectly preserved spiderweb specimen. The various parts of the web can be examined in more detail with a magnifying glass.

Explanation

Over the course of millions of years, female spiders have developed various kinds of web designs, suited for each spider's environment and the kind of food it likes to eat. Spiderwebs have various parts, such as the foundation, or anchor threads, which the spider spins first. The spider then fills in the rest of

Spider Web Varieties

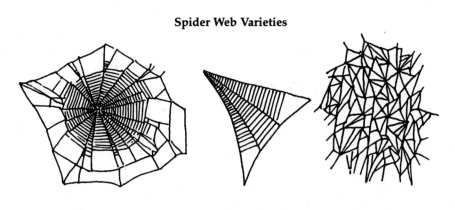

Orb Web   Triangle Web   Cobweb

the web by swinging between foundation threads.

The silk of a spiderweb is produced by specialized organs in the spider's abdomen called *spinnerets*.

Display Tip
Mount each preserved web specimen in a glass frame, and provide information about where you found it.

Did You Know?
When scientists wanted to know how well spiders could build webs in a weightless environment, they sent some spiders along with the astronauts. The astronauts found that, after a few moments of confusion, the spiders started building their webs as perfectly as they do on Earth.

Bird Pudding

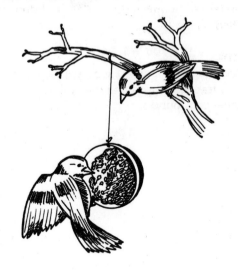

You Will Need

- Shallow plastic bowl (that you don't want to use again)
- Collected grease from bacon, hamburger, or any other meat
- Scraps of apples, bread; raisins, nuts, or sunflower seeds
- 2 reusable bowls, one smaller than the other
- Tablespoon (15-ml spoon)
- Warm water
- Strong cord
- Scissors

You can make a tasty pudding for birds, although *you* won't want any. Your bird pudding will be a real treat for hungry birds during the wintertime when plant and bug food is scarce.

Procedure

1. Half-fill the larger reusable bowl with very warm, but not hot, water.
2. Place the collected fat into the smaller reusable bowl and add the food scraps.
3. Place the smaller bowl into the larger bowl so that the fat begins to soften.
4. Stir the food scraps through the melting fat, then remove the bowl.
5. Use the scissors to cut four evenly-spaced slits around the edge of the plastic bowl.
6. Cut four lengths of cord and tie knots at the end of each.
7. Push the knotted ends of the cords through the slits on the plate so that the plate can be held up by the cords.
8. Knot the four cords together at their opposite ends.
9. Spoon the cooled bird pudding into the shallow bowl, and spread it flat.
10. Attach a long piece of cord to where the four cords knot together.
11. Hang your bird-pudding feeder on a tree.

Result: The design of your bird pudding feeder makes it easy for many kinds of birds to perch and feed. Acrobatic sparrows may land the edges of the bowl or on the cords above, while larger magpies will probably sit right in the center. You may watch quite a few birds fight and compete for food, but a regular refilling of your feeder should keep all the birds happy.

Explanation

Animals crave fat during the cold winter months to give them the energy they need. Birds are no exception. Since fat has many calories and calories mean energy, small portions of fat provide as much energy as larger portions of protein—now scarce. Observe your bird-pudding feeder at different times of day to see what kinds of birds feed from it.

Display Tip

Write down your favorite recipe for bird pudding, and have someone take pictures when you make the pudding. Take pictures when you build the feeder, too, and don't forget to show the birds enjoying their meals. If you'd like, clean your feeder (or make a new one), and display it in your booth next to all the photographs.

A Bug's Favorite Color

You Will Need

- Red, yellow, green, blue, and purple sheets of construction paper
- White sheet of paper
- Pencil or marking pen
- Wristwatch
- Ruler
- Clipboard
- Honey

Insects are hardly color-blind. In fact, some of them even have favorite colors. These colors remind them of flowers that contain their favorite nectars. In this project you can find out what colors attract certain insects—and then trick them.

Procedure

1. Draw the chart shown below on the white sheet of paper. You can make your chart wider by adding more columns of insects, if you'd like. Or, you can add one column for unknown bugs.
2. Make a copy of the chart.
3. Attach both charts to a clipboard, one chart covering the other.
4. Place the colored pieces of construction paper on a flat, grassy surface. Put a few stones around the edges of the paper to keep them from moving or blowing away.
5. Sit no more than 3 feet (90 cm) from the construction papers with your clipboard. Keep very still and be patient. Insects will begin to land on the different colors.
6. Make a slash in the appropriate box of your first chart every time a particular type of bug lands on a certain color.
7. After 20 minutes, add the number of slashes in your boxes to see what kinds of insects were attracted to the colors.
8. Remove the first chart from your clipboard, uncovering the second chart.
9. Place a dab of honey on the least popular color of construction paper.
10. Wait another 20 minutes and record your results as before.
11. Compare the results of the two charts.

BUG / COLOR	FLY	BEE	MOTH	GNAT	BUTTERFLY	?
RED						
YELLOW						
GREEN						
BLUE						
PURPLE						

Color Preference Chart for Bugs (steps 1–2)

Results: The red and purple sheets of construction paper attract larger insects like moths, butterflies, and some bees, while the blue sheet attracts more flies. Watch out, though, the yellow attracts "yellow-jacket" bees, as anyone who's ever worn bright yellow to a picnic knows! The least popular color is green, since green flowers are rare. But when you place a dab of honey on the green sheet, the bugs come to it, ignoring their original color preference.

Explanation

Insects prefer the colors they associate with their favorite nectars. And a favorite color can develop in response to an insect's environment. If you find, for example, bees swarming all over the bright yellow of a squash blossom, remove the blossom. The bees will still look for yellow for awhile, until they discover the succulent clover flowers nearby. Then they'll look for purple. Placing a dab of honey on the green paper caused the insects to ignore color and follow scent, since scent means the certainty of food. This is why, no matter how colorful your surrounding, insects will swarm around your very uncolorful sandwich at a picnic.

Display Tip

Document your color experiment with photographs and display your charts. Based on those results, collect the kinds of flowers you think the insects in your area would be most attracted to.

Did You Know?

Most insects see colors invisible to us. Colors of very short wavelengths, like ultraviolet, allow many insects to see the Sun through a thick layer of clouds and continue to navigate. The centers of many flowers also have something called *pollen lines*. These lines, sometimes visible to human eyes and sometimes not, actually point to the portion of the flower that contains the nectar. A flower *wants* an insect visitor to find nectar because, as in the case of bees, the insect also collects pollen on its body to fertilize other plants.

Bug / Color	FLY	BEE	MOTH	GNAT	BUTTERFLY	?
RED		II	III		IIII	
YELLOW		IIII IIII II		III		
GREEN	II	I		I		
BLUE	IIII III					
PURPLE	I	II	IIII I		III	

Completed Color Preference Chart

Creative Concoctions

Homemade Perfume
Mysterious Membranes
Sugar-Shriveled Egg
Sucking-Salt Solution
Cave Icicles
Shooting Volcano
Lava-Flow Volcano
Strange Flowing Rock
A Bottled Tornado
Natural Preservatives
Creamy Plastic
Dissolving Leaves
Make Your Own Paper
Immiscible Liquids
Oxygen Scale
A Test for Marble

Homemade Perfume

You Will Need

- 7 small jars or vials with lids
- Rubbing alcohol
- Popsicle stick
- Cotton swab
- Tweezers
- Paper towel
- ⅛-cup sample of each: fragrant rose petals, gardenia blossoms, orange-tree (or lemon-tree) leaves, eucalyptus leaves, pine needles, mint leaves, and whole cloves
- Tape and marking pen for labeling

Besides enhancing the flavors of foods, chemistry can please our sense of smell, too. For thousands of years, people have collected the aromatic oils of plants and seeds in order to make sweet-smelling waters and perfumes. But they had to squeeze a great number of plants to get just a few drops of oil, and the scent of the oil did not last very long.

Perfume-making is probably one of the oldest forms of chemistry. Perfume makers soon realized that, by adding other ingredients to a plant's essential oils, not only could less oil be used, but the essential oil's scent would last longer.

Procedure

1. Press as many plant and flower samples into ⅛ cup (30 ml) as you can. Place each sample in its own jar.
2. Except for the cloves, crush the samples as finely as you can with the Popsicle stick.
3. Add 2 teaspoons (10 ml) of rubbing alcohol and continue crushing.
4. Add about 10 cloves to one of the jars and then add alcohol.
5. Puts lids on all the jars, and allow them to sit in a warm place for about a week.
6. After a week, open one of the jars and dip in the cotton swab. Lift the swab toward your face, and fan the air around the moist tip so that the odor reaches your nose.
7. Dab the moist tip against the back of your wrist, then allow the spot to dry. Smell it.
8. Use the tweezers to remove a sample of the plant material, and let it dry on the paper towel. Smell it.

Result: The moist swab had a strong alcohol scent mixed with the plant scent. After you allowed the liquid to dry on your skin, your skin had only the plant scent and no alcohol odor. The sample of dried plant has little or no scent.

Explanation

Alcohol dissolves the aromatic oils in plants so that the plant's oils are removed from the plant tissue, suspended in the alcohol, and preserved. Alcohol also evaporates very quickly when exposed to air. When you placed a sample of homemade perfume on your wrist and exposed it to the air, the alcohol dried quickly, leaving behind only the aromatic oil.

Display Tip

Document each stage of your perfume-making procedure with photographs. Place the actual jars of perfume on your display table, along with some clean cotton swabs so that your perfumes may be sampled. You can also dab a small amount of perfume on separate index cards and label them. Place a sample or picture of each type of plant you've used to create your perfumes.

Did You Know?

About a hundred years ago, perfume manufacturers used a secret ingredient as a fixative to keep the scent from evaporating too quickly. The ingredient, *ambergris,* is a waxy liquid that coats the stomachs of sperm whales and protects the whales from the sharp bones of cuttlefish. Ambergris has the strange property of turning into a solid as soon as it is removed from the whale and exposed to air, and early photographs of whale hunters show them covered with icicles of ambergris as they packed the stuff into pails for transporting. Luckily, no one uses ambergris anymore due to chemical substitutes, and the needless killing of whales has almost stopped.

Mysterious Membranes

You Will Need

- 1 medium-size jar
- Shallow bowl
- White vinegar
- Raw egg in its shell
- Tongs
- Paper towel

In living things, many surfaces that appear solid are actually *semipermeable* to allow certain substances to pass through them. The semipermeable membrane surrounding a cell lets the cell absorb nutrients and dispose of waste products. This process is important to all living organisms. This project demonstrates the inward and outward flow of materials through the semipermeable membrane of a raw egg.

You can combine this project with Sugar-Shriveled Egg.

Procedure

1. Use the tongs to carefully place the raw egg into a jar. If you accidentally crack the egg, discard it and use a new egg.
2. Cover the egg with white vinegar, and observe it for the next 20 minutes, noticing the tiny bubbles that appear on the eggshell.
3. Leave the egg in the vinegar for the next five days, observing it periodically. Record any changes you see in the egg's appearance.
4. After five days, carefully pour out the vinegar solution from the jar, and *carefully* remove the egg with the tongs, placing it on the paper towel.

Result: The eggshell has completely dissolved, and the egg, while still raw, has grown larger and feels rubbery.

Explanation

When you placed the egg in the vinegar, you changed the chemical composition of the egg. The eggshell is made from a substance called calcium carbonate. Vinegar dissolves the calcium carbonate of the shell and releases carbon dioxide gas—the reason for the bubbles.

After the eggshell was dissolved, the semipermeable membrane surrounding the egg white was forced to do its work. Reacting to the vinegar, the membrane became more rubbery as it allowed some vinegar to pass through into the egg's interior. Since the egg contains a higher concentration of dissolved materials than the vinegar solution surrounding it, the egg "pulls" the vinegar into itself through a process known as *osmosis*. This is why the egg appeared larger.

Display Tip

Because of the perishable quality of eggs, you probably won't want to exhibit your experiment after you've performed it at home. But make sure you document the experiment with photographs, and carefully list both the procedure you followed and your observations.

Sugar-Shriveled Egg

You Will Need

- 1 medium-size jar, with lid
- Rubberized egg (from Mysterious Membranes)
- Corn syrup
- Tongs
- Paper towel

Procedure

1. Place the rubberized egg from the Mysterious Membranes project (p. 63) in a jar, and fill the jar with enough corn syrup to just cover the egg.
2. Place the lid on the jar, and allow the jar to sit undisturbed for five days. Observe the egg and record your observations.
3. After the fifth day, carefully remove the corn syrup from the jar. Then remove the egg with the tongs and place it on a paper towel.

Result: The egg has shriveled to just a fraction of its original size.

Explanation

When you placed the egg in corn syrup, the egg shriveled. This is because the higher concentration of material in the syrup pulled much fluid from the egg, but left behind many of the larger protein and fat molecules. These molecules were too large to fit through the membrane, and this demonstrates again why scientists call it *semi*-permeable.

Display Tip

Since eggs are perishable, you probably won't want to exhibit your experiment after you've performed it at home. But make sure you document the experiment with photographs, and carefully list both the procedure you followed and your observations.

Did You Know?

Scientists consider an egg the largest living cell, and eggs are useful for all types of experiments involving cell metabolism.

Sucking-Salt Solution

You Will Need

- Potato slices
- Shallow bowl
- Salt
- Paper towel

As with eggs, the cells of a potato are semipermeable to allow certain substances to pass through them. This quality of semipermeability allows cells to absorb nutrients and to dispose of waste products. The process is important to all animals and plants. This project demonstrates the outward flow of materials through the semipermeable membrane of a potato.

Procedure

1. Have an adult help you slice a medium-size potato into about ten pieces.
2. Place warm water in the shallow bowl, and add enough salt to make the water extremely salty to the taste.
3. Add the sliced potatoes, and leave them in the solution overnight.
4. Remove the potatoes from the salt water, place them on the paper towel, and examine them.

Result: The potatoes feel limp and rubbery.

Explanation

The higher concentration of dissolved material (salt) in the saltwater pulled the nonsaltwater out of the slices. Since the potato cells are held up by water, removing the water, or *dehydrating* the cells, caused the potato slices to lose their *turgidity* and become limp. The result of the osmosis through potato-cell membranes resulted in limp and rubbery potato slices.

Cave Icicles

You Will Need

- Sheet of black poster board
- Tape
- Scissors
- 4 drinking glasses, same size
- Black yarn
- Cellophane tape
- 4 washers
- Ruler
- Epsom salt
- Small saucepan
- Pot holder
- Tablespoon
- Water

Allow yourself at least two weeks to prepare this project, and don't expect dramatic results unless you live in a humid or damp climate. But if conditions are right, you can create a living cave—one in which *stalactites* and *stalagmites* grow. These "cave icicles" form when mineral-rich water flows through porous rock, leaving deposits behind.

Part 1 Constructing the Cave

Procedure

1. Cut a 4½ × 9-inch (11.25 × 22.5-cm) piece from the black poster board, and bend the piece into a semicircle.
2. Place the four drinking glasses around the outside of the semicircle, following the illustration. Tape the ends of the semicircle to the outside glasses. Check to make sure that the tops of the drinking glasses are at least 1 inch (2.5 cm) above the top edge of the poster board. If not, you'll need to either trim the poster board or find taller glasses.
3. Cut the black yarn into two 16-inch (40-cm) pieces. Tie washers to the ends of each piece of yarn.
4. Stretch the yarn pieces across the glasses so that the pieces crisscross and sag slightly.

Part 2 Mixing the Mineral Solution

Procedure

1. Fill the small saucepan with water, and add Epsom salt, stirring with the tablespoon until no more salt dissolves.
2. Heat the water (don't boil it) and continue stirring. You will find that you can add even more Epsom salt.
3. Have an adult help you carefully fill each glass with the warm Epsom-salt solution. Use the pot holder for this.

Caution: To avoid breaking the glasses, slowly pour the warm solution into each glass.

4. Allow your cave to sit for about a week, and observe any changes.

Result: The Epsom-salt solution flows into the draped yarn and accumulates at the point where the yarn hangs the lowest. As the solution drips from the yarn and dries, long stalactites form. These stalactites also create icicles under them—stalagmites—so that in about a week or two, you will have a very realistic-looking cave.

Explanation

The formation of stalactites and stalagmites really does resemble the formation of icicles. Through *capillary action*—or the force of attraction between a solid and a liquid—mineral-rich water flows through porous

Cave Icicles

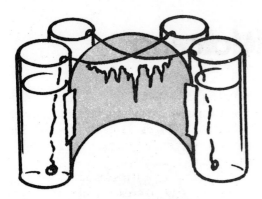

Formation of "Stalactites"

rock and accumulates. Eventually, gravity forces the accumulated water to drip out of the rock. As the droplets hang, the air evaporates some of the water, so that a mineral nucleus forms on which more droplets can hang.

In the case of icicles, super-cooled water drips though the icicle and freezes around a nucleus of ice at the tip. Droplet accumulates on droplet until a stalactite forms. Eventually, the stalactite grows large enough so that water can flow through it—again, through capillary action—right to the tip. Some heavier droplets fall from the tip of the stalactite to the ground, forming the upside-down icicle called a stalagmite. But here, evaporating water droplets pile up and flow down the stalagmite, leaving behind a growing accumulation of minerals.

Display Tip

Make sure you give yourself at least two weeks to build your cave and allow the cave icicles to form. Document the construction of your model with photographs and, if you want, determine the daily growth rate of your stalactites and stalagmites by carefully holding a ruler to them and recording your results. Your finished model makes an impressive display, particularly when you can clearly explain the process of stalactite and stalagmite formation through drawings and diagrams.

Did You Know?

Here's a memory trick (mnemonic) for remembering that stalactites hang and that stalagmites grow from the floor. The third syllable of *stalactites* begins with a *T*—for "top." And by the way, you don't have to visit a cave to see stalactites, they form on all sorts of porous stones structures where water can pass through, including bridges and buildings.

Shooting Volcano

You Will Need

- 2 liquid dishwashing detergent bottles, one with cap
- Tablespoon
- Red food coloring
- Vinegar
- Baking soda
- Papier-mâché
- Stiff cardboard or piece of wood
- Masking tape
- Brown and black paint
- Paintbrush
- Spray varnish
- White glue
- Funnel

A fiery, lava-spilling volcano is one of the most beautiful and terrifying sights in nature, and scientists who visit active volcanoes sometimes find themselves in great danger. This volcano model will allow you to sample some of the magic of volcanoes—safely at home.

You can combine this project with the Lava-Flow Volcano project.

Part 1 Shooting-Volcano Construction

Procedure

1. Fill one of the bottles three-fourths full of vinegar. Use a funnel if necessary. Add red food coloring, and put the cap on the bottle. Label the bottle "lava."
2. Place the second bottle in the center of the cardboard or wood, attaching it with a little white glue.
3. Tear the masking tape into strips, and attach the strips so that you make a kind of tent around the bottle.
4. Make papier-mâché by mixing flour and water in a bowl into a thick paste. Dip newspaper strips into the paste, and cover the tent you made around the bottle with the paste-coated newspapers. Build a little papier-mâché up around the lip of the bottle so that you form a crater.
5. Allow your volcano model to dry. Paint it brown and black to look like a mountain, and coat it with spray varnish.

Part 2 Shooting-Volcano Eruption

Procedure

1. Remove the cap of the "lava" bottle, and carefully pour the "lava" into the volcano bottle, using the funnel if necessary.
2. Quickly add 4 tablespoons (60 ml) of baking soda.
3. Stand back and watch your volcano erupt.

Result: The alkaline baking soda reacts with

the acid vinegar to produce carbon-dioxide foam. As the foam rises to the narrowing top of the jar, it gains momentum and shoots out to fill the crater of the volcano.

Explanation
The pressure inside a volcano can either result in a tremendous explosion or in plumes of lava that shoot far out into the air. The behavior of the foam in your miniature volcano simulated the pressure of ejecting lava.

Display Tip
It's better to take photographs of your erupting volcano rather than to demonstrate it in your booth. You should certainly display your impressive model, however.

Did You Know?
The pressure of molten-lava gases that builds up inside a volcano can equal the force of several atomic explosions.

Millions of years ago, a volcano in the northwestern United States blew the top off an entire mountain. The result was a gigantic crater, called a *caldera*, that gradually filled up with water. Today, that extinct volcano and transformed mountain is known as Crater Lake. In the state of Oregon, Crater Lake attracts thousands of tourists each year. Crater Lake is over 5 miles across. It is one of the deepest lakes in the world and one of the highest, too.

Lava-Flow Volcano

You Will Need

- Modeling clay
- 2 small water-cooler cone-shape cups
- Measuring cup
- Tablespoon (15-ml spoon)
- White liquid dishwashing detergent
- Sand
- Red food coloring
- Stiff cardboard or piece of wood for a base
- Masking Tape
- Stapler

This project demonstrates the two types of lava flow associated with erupting volcanoes, one more destructive than the other. *You can combine this project with the Erupting Volcano project.*

Procedure

1. On the cardboard, use modeling clay to make a small volcano cone, leaving a small crater at the top.
2. Fold and staple one of the water-cooler cone-shape cups so that you have a very shallow cup.
3. Fit this shallow cup into the crater of your volcano.
4. In a measuring cup, mix ½ cup (120 ml) of dishwashing detergent with red food coloring.
5. Pour this into the top of your volcano until the crater overflows and the "lava" spills down the sides.
6. Remove the paper cup from the crater, and wipe the volcano clean. Rinse the measuring cup.
7. Mix ¼ cup (60 ml) of dishwashing detergent with ¼ cup (60 ml) of sand in the measuring cup. Add red food coloring.
8. Pour this into the crater as before, until the lava overflows and spills down the sides. Use the tablespoon (15-ml spoon) to spoon out remaining lava.
9. Compare the flow of sandy lava to soapy lava.

Result: The soapy lava flowed smoothly down the sides of the volcano while the sandy lava formed clumps that crumbled down the sides.

Explanation

The plain detergent demonstrated fluid lava flow, and the sand-and-detergent mixture showed how slightly cooler lava behaves as it slips down the volcano sides. Fluid lava is usually more destructive because it travels farther than the crumbly lava.

Did You Know?

In Hawaii, the natives have words for the two types of lava: *pahoehoe* and *aa* (pronounced "ah"). *Pahoehoe* is the shooting, crumbly kind of lava, and *aa* is the more fluid kind.

Strange Flowing Rock

You Will Need

- Plastic cup
- Small bowl
- Cornstarch
- Measuring cup
- Measuring spoons
- Stirrer
- Water

Geologists know that the great land masses of Earth are continually drifting about the Earth's surface. This is happening even though the Earth's outermost layer, called the *crust*, appears solid or brittle, and the layer below that, called the *mantle*, is made of solid rock. Yet, near the top of this mantle is a zone that is not as solid as the rest. Called the *zone of partial melting*, the rock here behaves like a very strange liquid.

Although you can't recreate the actual type of rock found in the zone of partial melting, you can imitate the way it behaves with a strange puttylike mixture you'll make in this project.

Procedure

1. Pour ½ cup (120 ml) of water into the plastic cup, and gradually mix in 1 tablespoon (15 ml) of cornstarch.
2. Stir the water thoroughly until the cornstarch completely dissolves.
3. Add 19 more tablespoons (285 ml) of cornstarch (gradually, to avoid lumping), stirring with each addition, until the cornstarch and water mixture becomes very hard to stir.
4. Pour the mixture into the small bowl, observing how the mixture behaves as it flows.
5. Pick the putty rock up and play with it. Notice how it melts in your hand but also becomes very hard when you compress it.
6. Place the putty rock back into the bowl and allow it to settle. Then pick it up and bend it suddenly. It breaks!

Result: The putty rock behaves both like a rock and like a liquid.

Explanation

Scientists use the word *viscosity* to help describe the behavior of flowing substances. Viscosity means the internal friction of a liquid which affects the rate of its flow. For example, maple syrup has a high viscosity while alcohol has a low viscosity. In the zone of partial melting, high pressure turns rock into a partially liquefied substance of high viscosity—a kind of molten sea upon which the continents drift.

Display Tip

Take photographs to document how you made the strange flowing rock. Display the rock in a bowl for the judges to observe directly. Make a drawing of the Earth's interior and identify the zone of partial melting, where putty rock is found.

Did You Know?

Even though glass seems like a solid substance, it's actually a liquid of very high viscosity. The molecules of glass are arranged in long chains, much like the molecules of water. Very old panes of glass are actually thicker at the bottom than at the top—signs of flowing! Scientists have a special name for this kind of extremely high-viscosity material: *super-cooled liquid*. Over time, all rock flows. A marble bench in an old graveyard may have sagged, but that's not because it is worn, but because the marble flows.

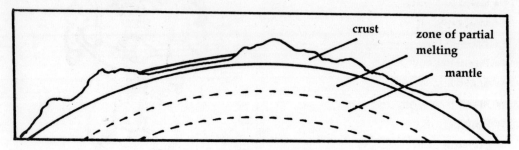

Earth's Interior

A Bottled Tornado

You Will Need

- Drinking glass
- Vegetable oil
- Tube of brown oil paint
- Funnel
- Nail
- Small fine-tooth saw
- Epoxy glue
- Popsicle stick
- Washer (slightly larger in diameter than the bottle's mouth)
- 2 clear plastic water bottles (same size)
- Duct tape
- Pepper
- Paper towel

Tornadoes terrify and fascinate us, and scientists have only begun to understand the forces that create them. By knowing how tornadoes begin, someday we'll be able to better predict when and where they'll happen. Observing tornadoes in action also reveals interesting information about how they move. Although this project is more whirlpool than tornado, it still reproduces the powerful vortex of tornado motion.

Part 1 Tornado Mixture

Procedure

1. To make the tornado vortex more visible, mix vegetable oil and water. Pour some vegetable oil in the drinking glass, and add a little brown oil paint to darken it.
2. Mix the paint and oil together with the Popsicle stick. Add about 1 teaspoon (5 ml) of pepper to this oil mixture. The pepper will really help your funnel stand out by tracing its twisting motion.
3. Remove the caps from the clear plastic water bottles and empty bottle #1.
4. Carefully saw a groove in the neck of bottle #1 so that you can widen the neck a bit to insert the washer. But don't worry about the washer for now; first you have to fill the bottle with the tornado mixture.
5. Insert the funnel and add water from bottle #2 until bottle #1 is about three-fifths full. Fill another fifth with the colored oil-and-pepper mixture, and leave the last fifth empty for air. Pour out the water remaining in bottle #2.
6. Now it is time to insert the washer into bottle #1. Spread the neck apart, and push the washer as far down as you can without allowing it to drop into the bottle.
7. Turn bottle #2 upside down, and insert its neck into the spread neck of bottle #1. The two bottles should fit together snugly, joined at the necks and separated by the washer.
8. To ensure that this joint stays watertight, spread epoxy glue around the joined bottle necks, and allow the glue to dry

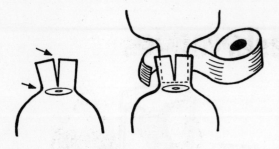

Secure bottles atop each other. (part 1, step 8)

overnight. Finish the joint by wrapping duct tape around it. (See illustration above.)

Part 2 Tornado in a Bottle

Procedure
1. Flip the bottles upside down so that the filled bottle sits at the top and the empty bottle rests on a table.
2. Support the lower bottle with one hand, and place the other hand on the upper bottle.
3. Quickly swirl the two bottles several times to start the vortex. Observe what happens inside the bottle.

Result: The brownish oil snakes toward the bottom of the bottle in a distinctive funnel shape, dragging particles of pepper "debris." Notice how the pepper swirls faster as it reaches the bottom of your tornado.

Explanation
Like a real tornado, your whirlpool tornado is shaped by powerful forces. But unlike your bottled version, real tornadoes occur when warm and cool air fronts collide suddenly. This collision leads to two layers of air with greatly different densities and pressures. As warmer moist air struggles to rise through cooler air, a powerful *convection current* begins to rotate upward and accelerate. Crosswinds put a spin on this convection current, resulting in a whirling column of upward-moving air called a *mesocyclone.*

As a tornado starts to take shape, the narrow part of the funnel spins more rapidly than the wider part of the funnel. This increased angular momentum at the tip draws the tornado down like a wedge, until it touches the ground.

Display Tip
Document the construction of your bottled tornado with photographs. Exhibit your finished tornado and demonstrate it. Explain how your bottled tornado resembles a real tornado and how it differs. Collect photographs of real tornadoes and display them.

Did You Know?
Scientists have begun to suspect that another type of tornado exists—a horizontal tornado that's mostly invisible to the naked eye. Called *rotor-vortex microbursts*, these tornadoes can appear suddenly and create hazardous conditions for low-flying airplanes. Rotor-vortex microbursts seem to occur in flat areas surrounded by mountains, but scientists still have much to learn about them.

Natural Preservatives

You Will Need

- 1½-ounce (42-g) beef-bouillon cube
- Bowl
- 2 cups (480 ml) of warm water
- 4 small plastic drinking glasses, same size
- Measuring cup
- Measuring spoons
- White vinegar
- Salt
- Lemon
- Waterproof marker

Before refrigeration, people had to find ways to keep foods from spoiling. For some kinds of meat and fish, drying was the answer. This also worked for many vegetables and fruits. But liquid foods needed something to delay the growth of bacteria. These ingredients had to be edible and even make the food taste better, if possible. You can experiment with some of these *natural preservatives* in this project.

Procedure

1. In a bowl, dissolve the beef bouillon in 2 cups (480 ml) of warm tap water.
2. Use the waterproof marker to label the plastic glasses: S (salt), V (vinegar), L (lemon), and C (control).
3. Add ½ cup (120 ml) of broth to each of the four glasses.
4. Cut the lemon in half.
5. Mix 2 teaspoons (10 ml) of salt into the S glass, 3 teaspoons (15 ml) of vinegar into the V glass, and 3 teaspoons (15 ml) of lemon juice into the L glass (vinegar and lemon juice are liquids and less concentrated than the salt). Don't add anything to the control glass—it's the glass against which you'll measure all other results.
6. Place the glasses in a warm spot, and leave them undisturbed for about a week.

Result: All glasses become cloudy, with the control glass the cloudiest of all. The glass containing the lemon (L) is slightly less cloudy than the control glass (C), and the glass containing the salt (S) is less cloudy than the glass containing the vinegar (V). Of all the glasses, the glass containing the vinegar (V) is the least cloudy.

Explanation

Vinegar has the best preservative powers, followed by the salt and lemon juice. Broth, without any preservative at all, spoils quickly.

Of the three, lemon makes the poorest preservative because lemon juice is a food, and when it loses its acidity, it attracts bacteria. Vinegar not only works as a preservative but it tastes good and contains vitamins and minerals. Do you think this might have something to do with the invention of pickled foods?

Display Tip

Most science-fair regulations do not allow the display of cultures containing molds or bacteria. You'll need to carefully document the progress of your experiment with photographs. Try experimenting with other foods, such as sweet foods and various acidic or salty preservatives. By trial and error, see if you can find out what preservative works best for which food. Remember, for a preservative to be truly useful, it must also be NONTOXIC!

Caution: Never taste your samples. Just watch them for cloudiness.

Creamy Plastic

You Will Need

- ½ cup (120 ml) of light cream or half-and-half
- 1 teaspoon (5 ml) of white vinegar
- Nonstick saucepan
- Pot holder
- 2 coffee filters
- Rubber band
- Coffee cup
- Paper towel
- Aluminum foil
- Plastic button
- 2 clay flowerpots

This project helps you see that difference for yourself, but you may need an adult's help for the heating and pouring part.

Part 1 Button-Making

Plastic made from petroleum is only a recent invention. During the 19th century, many plasticlike products were made naturally from animal and plant material. On some of the barns of New England, for example, you can still see traces of 100-year-old weatherproofing in the form of milk "paint." Sometimes, the milk was mixed with livestock blood to make it even more resistant to the weather.

Other organic plastics had some very strange properties. A plastic made from the by-products of cotton milling was very popular in the 1870s. It was widely used to make billiard balls until someone discovered that the plastic had combustible properties and could explode when hit hard enough. Many a saloon gunfight might have started with a gunlike crack during a game of pool.

One of the important differences between natural plastics and today's petroleum product is in the degree of decomposition. Plastics made from plants or animals will dissolve in a landfill, while many petroleum products never do.

Procedure

1. Pour ½ cup (120 ml) of light cream into a saucepan and heat until the cream begins to foam slightly. Remove the cream from the heat.

2. Add 1 teaspoon (5 ml) of white vinegar to the cream and stir. The cream will begin to clump into tiny curds resembling farina, surrounded by a clear liquid. If you don't see any clumping, add a little more vinegar.

3. Place two coffee filters together and tuck them into the coffee cup. Secure the filters to the cup with a rubber band.

4. Using the pot holder, carefully pour the clumping cream mixture into the coffee filters. Make sure you scrape the pan to get all the curds out.

5. Allow about 5 minutes for the curds to cool, then lift the coffee filters from the cup, wrap them around the plastic, and squeeze out any excess liquid.

6. Unwrap the filters. You should have a pure white, cheeselike plastic that can be easily molded.

7. On a piece of aluminum foil, shape your plastic into several small buttons. Place the buttons on a paper towel to dry.

Result: In about 24 hours, your hand-molded buttons have turned into a hard, yellowish material—natural plastic! Compare these buttons to the ordinary plastic button.

Part 2 Burying the Evidence

Procedure
1. After the buttons dry, divide them so that you have several to bury and at least one that you'll keep for your display.
2. Take your buttons outside, along with the two flowerpots.
3. Fill each pot about halfway to the top with soil.
4. Place a few of your homemade plastic buttons in pot #1, and place the ordinary button in pot #2.
5. Cover the buttons with soil, and water the soil in the flowerpots every day for about a week.
6. Uncover the buttons in both pots and compare them.

Result: The homemade buttons crumble easily and fall apart. The ordinary plastic button shows no sign of decay.

Explanation
The cheeselike mass at the end of the curdling process consisted of fat, minerals, and the protein *casein*, a stringy molecule that bends like rubber until it hardens. All three of these substances combine to form a strong material that can still be broken down when exposed to wet soil and changing temperatures.

Display Tip
Document each stage of your plastic-making with photographs, and display samples of your creamy plastic buttons before and after they begin to decompose.

Did You Know?
Recently, scientists all over the world have developed different kinds of biodegradable plastics by adding chemicals that allow the plastic to break down when exposed to light, water, or bacteria. For example, a special plastic designed to dissolve in saltwater keeps plastic trash from harming sea creatures.

The Japanese have recently developed an inexpensive biodegradable plastic made from the shells of shrimp. Japanese scientists have figured out how to remove the material *chitin* from the shrimp shells and combine it with silicon. (Chitin is also the material in your fingernails.) The result, *chitisand*, is stronger than petroleum plastic. It also makes an excellent fertilizer when it breaks down in the soil.

Dissolving Leaves

You Will Need

- Several broad-leaf samples, such as ivy, maple, and oak
- Several narrow-leaf samples such as spiderplant and lily.
- Shallow bowl
- Baking soda
- Paper towel
- Bleach
- Tweezers
- Wax paper
- Heavy book
- Washers
- Spray varnish
- Black construction paper

To clearly see the vein structure, or *venation*, of leaves and flower petals, you can use a simple baking-soda solution to dissolve the pulpy plant tissue and turn your specimens into delicate and beautiful lacework. Since it takes about 2 weeks to get the best results, plan ahead.

Procedure

1. In a shallow pan, mix 1 teaspoon (5 ml) of baking soda with 2 cups (480 ml) of warm water.
2. Add broad and narrow leaves, making sure they sink to the bottom of the pan. If the leaves float, place a small washer on each one.
3. Put the pan in a sunny place and leave it undisturbed for about 2 weeks.
4. Use the tweezers to carefully remove the leaves, and rinse the leaves in a bowl of clean water.
5. Place the leaves on a paper towel to dry; then move them to a sheet of wax paper. Cover them with another sheet of wax paper and place a book on top. After a few days, remove the book and wax paper and examine your leaves.
6. Mix ½ cup (120 ml) of bleach with 2 cups (240 ml) of water, and pour the mixture into a shallow bowl as before.
7. Carefully lower each leaf into the bowl and submerge it. When the leaf whitens, remove it and place it on a paper towel to dry.
8. Mount your "skeletonized" leaves on a piece of black construction paper to bring out the detail.

Result: This process results in the clean removal of soft plant tissue, clearly exposing two beautiful venation patterns.

Explanation

Scientists divide plants into two subclasses, *monocots* (monocotyledons) and *dicots* (dicotyledons). A monocot plant is the older species and can be easily recognized by its structure of parallel veins. Plants like gladiolas, lilies, grasses, palm trees, and banana trees, as well as many evergreen trees and bushes, all have *parallel venation*.

Dicots evolved later and have a more sophisticated *branched venation*. Most of the familiar shade trees like maple, oak, chestnut, and elm, are all dicots.

By "skeletonizing" samples of leaves, you can more easily recognize the difference between monocots and dicots. Do you find any other distinctive characteristics in your leaf samples?

Display Tip
Place each skeletonized leaf sample on a piece of black construction paper and gently cover each with a light coating of spray varnish.

Make Your Own Paper

You Will Need

- Stack of old newspapers
- Large soup pot
- Small picture frame (must fit inside soup pot)
- Small piece of screen or fine netting
- Thumbtacks
- Wooden spoon
- Old blanket or piece of felt
- Plywood board

You can easily make your own paper with a few kitchen and hardware-store materials. Since this project uses old newspapers, your papermaking is actually paper recycling. Most professional papermakers now use about 30-percent recycled paper for making new paper. Most paper is still made from soft-pulp trees that must be cut down and sawed into bolts for processing.

Procedure

1. Place a stack of old newspapers in the large soup pot and fill the pot with cool water. Allow the newspapers to soak for about 5 minutes; then pour off the water. This removes the loose printing ink.
2. Shred the wet newspaper in the pot. The fineness of your paper will depend on the size of your shredded pieces, with smaller pieces making a smoother and stronger paper.
3. Add water to the pot so that the paper is completely covered.
4. Allow the paper to soak overnight.
5. Ask for an adult's help for this step. Boil the newspaper and water until the newspaper dissolves into a kind of thick oatmeal. At this point, you can add some food coloring or pieces of dried flowers to customize your paper. Professional papermakers would add bleach to whiten the paper.
6. Allow the mixture to cool for several hours. A smooth and even layer of paper pulp will begin to form at the surface.
7. Stretch the piece of screen or netting over the picture frame and tack it around the edges.
8. On a flat surface, open the blanket (or piece of felt) and place the plywood board close to it.
9. Carefully dip the picture frame into the soup pot, and lift it out flat so that the mushy pulp forms a sheet on the screen. If you see holes in the sheet, spoon some pulp over them.
10. Let the water drain away. Then move to the blanket and quickly turn the frame over so that the sheet falls onto the blanket.
11. Fold the blanket over the sheet; then place the plywood board over the folded blanket.
12. Press down on the plywood to squeeze excess water from the paper.
13. Lift the board and unfold the blanket. Let the paper dry overnight.
14. When dry, carefully peel your paper from the blanket, and trim away the rough edges.

Result: You will have a rough but sturdy paper that you can use for greeting cards and artwork. Your paper may not fold without breaking, though; so test some of the paper before you use it for letters.

Explanation
This project actually recycled old newspapers. Paper consists mostly of wood pulp or *lignum*. When soaked, the pulp breaks apart into tiny fibers that can be mixed around in a kind of paper "soup," but they lock together again when the water is removed. Professional papermakers add other ingredients to paper—like clays and starches—to strengthen the locking of fibers. These bonding agents make the paper stronger, more flexible, and easier to fold without crumbling.

Display Tip
Record each step of your papermaking with photographs. Try recycling different kinds of paper, or add various ingredients to your paper stew—such as pieces of colored construction paper—for interesting results. Display your homemade product, comparing it with professional paper.

Did You Know?
One 35- to 40-foot (10.5- to 12-m) tree produces a stack of newspapers 4 feet (1.2 m) thick; therefore, it takes this much recycled newspaper to save one tree. One ton (0.9 metric ton) of recycled paper spares 17 trees and saves more than 3 cubic yards (2.28 cubic m) of landfill space.

Immiscible Liquids

You Will Need

- 3 small jars with lids
- Water
- Green food coloring
- Cooking oil
- Alcohol
- Liquid dishwashing detergent

Just because a liquid is "wet," it can still surprise you by refusing to mix with other liquids. When two liquids refuse to combine, we call them *immiscible*. This project demonstrates some of the strange behavior of immiscible liquids.

Procedure

1. Fill the first jar one-third full with water
2. Tip the jar slightly and add the same amount of oil by pouring it against the side of the jar.
3. Repeat this procedure for the alcohol.
4. Watch what happens to the three liquids in the jar.
5. Repeat steps 1–4 in the second and third jars.
6. Add a few drops of liquid detergent to the third jar, screw on the lid, and shake the jar vigorously.
7. Screw the lid on the second jar and shake it. (No detergent added.)
8. Arrange the jars in order 1–3, wait a few hours, and observe what happens to the liquids inside the jars.

Result: In the first jar, the water, oil, and alcohol form three distinct layers and do not mix. When you repeat this procedure in the second jar and shake it, the liquids appear to combine at first, but the layers re-form after a few hours. When you shake the soapy jar, the layers also combine but do not re-form.

Explanation

The three immiscible liquids—water, oil, and alcohol—have different molecular densities. The heaviest liquid is water, followed by the oil and alcohol. Shaking the second jar and waiting a few hours shows that these liquids do not combine. But the third jar shows that they *can* combine if detergent or soap is added. Detergent and soap belong to a class of substances scientists call *emulsifiers*. When you add soap to

Jar #1
Water, Oil & Alcohol

Jar #2 (shaken)
Water, Oil & Alcohol

Jar #3
Detergent Added

an oil-and-water mixture, soap molecules surround the droplets of oil to prevent them from coming together again.

Display Tip
Document your experiment with photographs, recording how your jars looked immediately after shaking and then after a few hours of waiting. Label the jars and display them for the judges.

Did You Know?
Chemists use emulsifiers to make our foods tastier. For example, egg yolk is used as an emulsifier in mayonnaise because it allows the oil to combine with the vinegar. And in ice cream, gelatin acts as an emulsifier to keep the cream from separating from the ice crystals, which helps keep your ice cream smooth.

Oxygen Scale

You Will Need

- Long, narrow strip of iron or steel (available in hardware stores)
- Small triangular dowel (available in lumber yards)
- Fine sandpaper
- Petroleum jelly
- Spray bottle for water
- Ruler
- Pencil

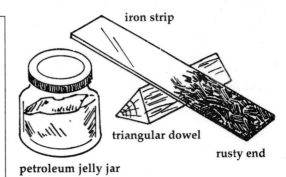

Oxygen Scale (step 6 and result)

What happens when a piece of steel or iron rusts? We all know what rust looks like, and that it begins to rust when the metals become wet. But what actually *is* rust? Is it just a changed form of iron or steel, or an entirely new material? This project will help you find out. Set it up near a windowsill, or even outdoors if weather permits. And make sure your strip is steel or iron and not stainless steel or galvanized iron. Neither will work.

Procedure

1. Clean the iron strip by rubbing it firmly with the sandpaper.
2. Measure the length of the strip, dividing it in half.
3. Coat half of the strip with a thin layer of petroleum jelly.
4. Carefully balance the strip on the triangular dowel.
5. Spray water on the uncoated half of the strip. The strip will dip with the weight of the water, then balance again as the water dries.
6. Allow the balanced strip to remain undisturbed for about a week.

Result: The uncoated half of the strip rusts while the coated half remain rust-free. But something else: the bar has tipped down toward the rusted side.

Explanation

Unprotected iron or steel reacts with moisture in the air. Moisture allows oxygen in the air to combine with the iron, producing rust. Rust is neither like the iron nor the oxygen that combine to create it, but a chemical combination of the two called *iron oxide* (Fe_2O_3). You can "see" this added oxygen in the way the strip tilts down toward the rust side.

Display Tip

Document the construction and operation of your oxygen scale with photographs. Display the model along with an explanation of what it demonstrates.

Did You Know?

Chemists sometimes add "heavier" elements to familiar compounds in order to study the results. In the case of "heavy water," they discovered that they could create a new form of water by removing the ordinary hydrogen in the water molecule (H_2O) and replacing it with a heavier form of hydrogen called *deuterium*.

The deuterium combines with oxygen perfectly (D_2O), and the result looks and tastes just like water, only it weighs more and has a higher boiling point.

Although not poisonous, no one should drink heavy water. Eventually, the deuterium begins to affect living tissue. In an experiment, plants fed heavy water for two weeks began to die, and animals placed on a heavy-water diet showed signs of disease.

A Test for Marble

You Will Need

- Some marble gravel, a piece of slate, and a piece of limestone (all available in gardening stores)
- White vinegar
- 3 small drinking glasses
- Tape and marker for labeling glasses

If you were a geologist hired to find a place that contained marble for quarrying, how would you proceed? In the rough, marble looks like many other metamorphic rocks, that is, rocks formed from other rocks. This project shows you how to test for marble. You'll also learn what other kind of stone is related to marble.

Procedure

1. With the tape and marker, label the glasses "marble," "slate," and "limestone."
2. Fill the "marble" glass half full with vinegar and add some marble gravel.
3. Observe the results.
4. Add vinegar to the "slate" glass and observe the results.
5. Add vinegar to the "limestone" glass.
6. Observe and compare what happens in all three glasses.

Result: Two of the three glasses bubble: the glass containing marble and the glass containing limestone. The glass containing slate does not react at all.

Explanation

Vinegar reacts with a substance found in both the marble and limestone: calcium carbonate. Vinegar, an acid, actually *dissolves* the calcium carbonate. With fresh supplies of vinegar, both the marble and limestone would eventually be eaten away. The slate contains no calcium carbonate and does not react to the vinegar.

Display Tip

Document your marble testing with photographs, and display the labeled glasses in your booth as a simulation of the experiment. Collect photos of marble buildings, and marble sculptures and other objects. Can you think of why these structures might be in danger?

Did You Know?

Marble is a chemically delicate material that can be easily damaged. Many substances can corrode marble, and these include the phenomenon we call *acid rain*. Not quite like ordinary rain, acid rain occurs when pollutants in the atmosphere combine with rainwater to produce a weak sulfuric-acid solution. This acid solution eats away and dissolves the marble. Many great marble structures have already been damaged by acid rain.

Marble is also easily damaged by fire. Although it doesn't burn directly, intense heat causes marble to turn into powder and sometimes even explode. This is why many of the great buildings of ancient cities were vulnerable to damage from devastating fires.

Find Out about Food

How Much Vitamin C?
Test for Fats
Looking for Sugar
Painted Apples
Artificial Sweeteners
Proteins in Food
Why Do We Cook Food?
The Mighty Starch Molecule
Iodine Test for Starch
Why Toast Tastes Better
Iron Content in Fruits & Vegetables
Iron Content in Breakfast Cereals
Bones & Minerals

How Much Vitamin C?

> **You Will Need**
>
> - Cornstarch
> - Small jar
> - Measuring cup
> - Measuring spoons
> - Stirrer
> - Saucepan (stainless steel or enamel, not aluminum)
> - 2 plastic sandwich bags that zip closed
> - Iodine
> - 250-mg vitamin C tablet
> - Small plastic cups
> - Eyedropper
> - ½ cup (120 ml) of each juice—orange, grape, apple, tomato, and green pepper

Vitamin C helps your body stay healthy. Not only does this vitamin help to repair tissues, but it aids in healing and may even prevent certain types of infections. Unlike dogs, human beings cannot produce vitamin C naturally, and so we need to eat vitamin C–rich foods every day. Many foods contain high concentrations of this important vitamin, and you'll test some of them in this project.

Procedure

1. Put ½ teaspoon (2.5 ml) of cornstarch in a saucepan with 1 cup (240 ml) of water. Make sure you use the measuring spoon for this, and make sure that you have a level, not heaping, spoonful.

2. Place the pan over low heat and stir until the cornstarch completely dissolves. Pour the solution into the jar and allow it to cool.

3. In a separate jar, mix 1 teaspoon (5 ml) of the starch solution with 1 cup of water (240 ml) and 4 drops of iodine. Adding iodine will turn the cornstarch-and-water solution blue, creating a *test standard* that you'll use to detect vitamin C by a process called *titration*.

Caution: BE VERY CAREFUL HANDLING THE IODINE. IT IS A POISON IF SWALLOWED, AND IT CAN STAIN YOUR SKIN AND CLOTHES.

4. To try out your test solution, dissolve the 250-mg vitamin C tablet in 1 cup (240 ml) of cold water. Prepare the vitamin for dissolving by first placing it in a plastic bag that zips closed and crushing it with a rolling pin or hammer.

5. Put 2 tablespoons (30 ml) of the test solution in a small plastic cup. Add 1 drop of the dissolved vitamin-C solution and stir. The blue color of the test solution disappears.

6. Place samples of the juices in separate plastic cups. Place the green pepper in a plastic bag that zips closed and crush it until you can extract the juice.

7. Repeat step #5, but substitute samples of each juice for the pure vitamin-C solution. Use a clean plastic cup and fresh testing solution for each juice test. Count how many drops of each juice you need to add before the blue color of the testing solution disappears. This drop-by-drop measuring technique is what is known as *titration*.

Result: It took fewer drops of the green pepper, orange, and tomato juice to remove the blue color from the testing solution and more drops of the grape and apple juice.

Explanation
The higher the concentration of vitamin C in a juice, the less of it you'll need to remove the blue color from the test solution. You made the test solution by adding iodine to cornstarch. The chemical bonding between these two substances created a new molecule that turned the solution blue. Adding vitamin C to the solution reverses this chemical bond and turns the test solution clear again.

The more vitamin C in a sample, the less of the juice you'll need to *undo the blue*.

Test for Fats

You Will Need

- Brown paper bag
- Food samples: butter, yogurt, whipped cream (canned), peanut butter, raw bacon, avocado, potato chip, chocolate-chip cookie
- Cotton swab
- Aluminum foil
- ½-teaspoon (2.5-ml) measuring spoon
- Tweezers
- Marking pen
- Scissors

This project tests and compares fat and oil content in everyday foods. Fats and oils contain more energy per pound than do other foods. For this reason, people living in cold climates depend on them for energy. Fats and oils resemble each other chemically, but they differ in their melting points. Fats are solid at room temperature, and oils are liquid at room temperature.

Procedure

1. Cut a large square from the brown paper bag. Make sure the square contains no printing.
2. Stretch out a large piece of aluminum foil to place your food samples on.
3. Place a ½ teaspoon (2.5 ml) of butter, yogurt, whipped cream, and peanut butter on the foil.
4. Cut the avocado in half and scoop out ½ teaspoon (2.5 ml) to add to your samples. Use the remaining avocado for dip. (Mash the avocado with a little lemon juice and ketchup, and get out the tortilla or potato chips.)
5. Slice a small piece of bacon and place it on the foil, followed with a few potato chips and a piece of chocolate-chip cookie.
6. Use the marking pen to make a list of your food samples on the square of brown paper. Leave lots of room between each item of your list.
7. For all the food samples, except the bacon, potato chips, and cookie, take a cotton swab and rub it in the sample.
8. Rub the cotton swab on the brown paper (next to the word on the list) to make a smear.
9. Pick up the raw bacon with the tweezers and rub it on the paper.
10. Pick up the chips and cookie and rub them on the paper.
11. Leave the paper for two days and observe it.

Result: Each food sample leaves a greasy spot on the brown paper, turning the paper a darker color. After two days, some of the greasy spots have spread out farther than others.

Explanation

The fats and oils in the food samples travel through the fibers of brown paper, leaving a dark brown smear. Some of the smear is just water, which will evaporate overnight. The

smears that spread the farthest after two days reveal the foods with the highest fat and oil content.

Display Tip

Document your procedure with photographs. Display your square with each smear identified. Test food samples other than those suggested. For an interesting variation you could test whole milk and nonfat milk, or regular cookies with nonfat varieties.

Did You Know?

Although many health and fitness experts recommend swimming for exercise, none recommend it for losing weight. Even if you burn many energy calories while swimming, your body will resist losing fat because the fat protects your body from the cool water temperature. This is why the fittest of otters, seals, and sea walruses have a thick layer of fat beneath their skins—to keep them warm.

Looking for Sugar

You Will Need

- 12 small plastic glasses
- Measuring spoon
- Measuring cup
- Paper-towel blotter
- Saucepan
- Vinegar
- Granulated sugar
- Reagent glucose strips (*visual diagnostic* type for urinalysis [urine analysis], available in drugstores)
- ⅛ cup (30 ml) of sample foods: apple, applesauce, corn, canned corn, tomato, bottled tomato sauce, ketchup, orange, lemon, milk, cola

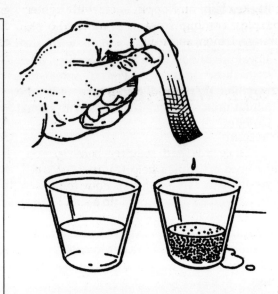

We know that many foods contain sugar naturally and that food manufacturers sometimes add sugar to food to enhance taste. But what kinds of foods contain natural sugar? And to what kinds of foods do manufacturers add sugar? This project tests a variety of foods for sugar by using glucose strips.

Part 1 Cool Testing

Procedure

1. Place ⅛ cup (30 ml) of all the food samples in separate glasses. Use the back of a spoon to mash the apple, corn and tomato, and thin the ketchup and applesauce by mixing in a teaspoon of water.
2. Place ⅛ cup (30 ml) of water in one of the remaining glasses, and stir in 2 teaspoons (10 ml) of granulated sugar. Fill the last glass with just water, your control for this test.
3. Place the glasses in this order: (1) apple, (2) applesauce, (3) corn, (4) canned corn, (5) tomato, (6) bottled tomato sauce, (7) ketchup, (8) orange, (9) lemon, (10) milk, (11) cola, (12) sugar water, and (13) water.
4. On index cards, identify the food samples and give each a number. Place each card next to its glass.
5. Spread a paper towel close to the food samples.
6. Dip a glucose strip in the first glass and remove it, scraping the excess liquid along the side of the glass. Place the strip on the paper towel and wait 10 seconds.
7. After 10 seconds, notice the color of the strip. The box containing the glucose strips may have a color chart. Compare the color of your strip with the chart. If not, write down the color you see.
8. Continue this procedure for the remaining food samples.

Result: Except for the plain water and sugar water, all of the foods turned the pink glucose strips a darker purple. Although both the apple and applesauce turned the strips purple, the applesauce purple was much darker. The canned corn was also a much darker purple than the fresh corn, as was the tomato sauce when compared with the tomato. The deepest purples came from the cola, ketchup, and applesauce, with lighter purples coming from the canned corn, orange, lemon, and milk.

Part 2 Warm Testing

Procedure

1. Since this part requires heating, you may need an adult's help (caution for children 12 years and younger). Pour the glass containing the sugar water into a small bowl and heat it to just below boiling.

2. Add about ½ teaspoon (2.5 ml) of vinegar and allow the sugar water to cool.

3. Pour the liquid back into the glass and retest with a glucose strip.

Caution: HOT SUGAR QUICKLY BURNS SKIN. DON'T TOUCH IT.

Result: The sugar water now reacts with the glucose strip, turning it a dark purple.

Explanation

All the food samples contain *simple sugars* that react to the glucose strips. Besides glucose (which includes the sweeteners corn syrup and dextrose), simple sugars consist of fructose (fruits), lactose (milk), and maltose (starch). Of the foods tested in this experiment, the milk contains lactose, and the apple, corn, tomato, orange, and lemon contain fructose. The applesauce, tomato sauce, and canned corn contain both fructose *and* glucose—or added sugar. The cola contains only glucose in the form of corn syrup.

But this isn't the whole story. *Sucrose*, or granulated sugar, is a *complex* sugar molecule that you must break apart before a glucose strip can detect it. Heating sucrose and combining it with an acid, such as vinegar, helps break it down.

Display Tip

An experiment such as this requires detailed documentation in the form of photographs. Exhibit rows of glucose strips comparing the sugar content of various foods. Expand your data by comparing fresh and store-bought fruit juices, canned meats, and low-fat desserts such as puddings or ice cream. You'll discover that low-fat or nonfat desserts contain much more added sugar than the regular varieties (and offer no real calorie savings).

Test lemons for sugar. You'll be surprised to learn that a lemon contains as much sugar, in the form of fructose, as an orange. In fact, it may even contain more sugar than an average-size grapefruit.

Did You Know?

Fructose is the sweetest of all sugars, followed by sucrose, glucose, lactose, and maltose. But these are only five of the hundreds of sugars scientists have discovered. Although certain sugars, such as cane sugar and maple sugar, may have a different taste, they're chemically the same. What gives many sugars their distinctive flavor is the impurity, or food particle, in the sugar.

Painted Apples

You Will Need

- 5 apples, from very green to very ripe
- Small plastic glass
- Iodine
- Cotton swab
- Aluminum foil
- Sharp knife

Many interesting chemical changes occur in fruits as they ripen. This project tests for chemical changes in ripening apples by painting them with an iodine solution.

Procedure

1. Have an adult take a thick slice out of the center of each apple. Don't try to slice the apples yourself.
2. Lay the apple slices side-by-side, from green to ripe, on the aluminum foil.
3. Add water to the plastic glass until it's half-full. Add 5 drops of iodine to the water and stir the cotton swab.
4. Using the cotton swab, paint the iodine and water solution over each apple slice. Make sure you cover each slice thoroughly.
5. Observe any changes in the apple slices.

Result: The unripe slice turns dark blue, except for the center section which remains white. The riper the slices turn less blue; the ripe slice remains the same.

Explanation

For fruits and vegetables, ripening means that starch turns into sugar. The less ripe the fruit, the more starch and less sugar it contains. Iodine causes a chemical reaction with starch that leads to the formation of a "blue" molecule, making the starch easy to see in your apple slices.

Display Tip

Since this is an easy experiment to prepare, mix up a fresh batch of apples to display in your booth. After you've brushed them with iodine, sprinkle a little lemon juice on them to keep the fruit from turning brown.

Artificial Sweeteners

You Will Need

- Artichoke
- Small pot with cover
- Knife
- Cutting board
- Plate and fork
- Glass of water
- Quinine water
- 2 cotton swabs
- Salt

Scientists study the flavors and odors of foods in order to imitate them chemically. For example, chemical *food enhancements* can make you taste fruit in a soft drink that contains no fruit, or sweetness in a desert that contains no sugar. Such enhancements can sometimes be very useful for people with dietary restrictions.

Procedure

1. First, have an adult help you prepare and cook the artichoke. Wash the artichoke by holding it under a stream of water. Cut off the stem, and place the artichoke in a saucepan.
2. Pour about 2 inches (5 cm) of water into the saucepan, cover it, and allow the artichoke to cook for about an hour, or until a fork pushes into it easily. Allow the artichoke to cool for about 30 minutes.
3. Pull off the leaves until you reach the soft center section, or heart. Scrape the tiny green leaves and hairy bristles away from the heart, and cut the heart into four pieces.
4. Take a sip of water. Then, take a piece of artichoke heart in your mouth and chew it very slowly. Take another sip of water, noting the taste.

Result: For most people, the second sip tastes sweeter than the first. But scientists believe that some people are genetically immune to the sweetening qualities of artichokes.

Explanation

When we taste something, taste buds on various areas of the tongue become stimulated. The brain receives this scattered information and interprets it as a single flavor, basing it on something we've already tasted, recognized, and either like or dislike. In this way, the brain learns to recognize flavors from a multitude of complex taste stimuli. And this is also how scientists learn to reproduce those flavors by analyzing the stimuli and reproducing them in the laboratory.

This is the case with aspartamine, the chemical found in artichoke hearts. Aspartamine stimulates the sweet-sensing areas of the tongue so that many foods, despite their complex flavors, now seem "sweet" to our brains.

Did You Know?

You can achieve the same artificial sweetening effect by using carbonated quinine water. You can buy quinine water, also called Tom Collins mix, in the soft-drink section of your supermarket. Pour a little into a paper cup, dip a cotton swab into it, and brush the solution on your tongue. A sip of water tastes sweet.

Salt affects the taste of water in a different way. Put 1 teaspoon (5 ml) of salt in a paper cup, and add just enough water to moisten the salt. Dab the salt solution on your tongue and then take a sip of water. The water now has a slightly sour and bitter flavor. Tricked again!

Proteins in Food

You Will Need

- Small saucepan
- Small disposable bowl
- Disposable eyedropper
- Measuring spoons
- 7 disposable plastic glasses
- Medium-size disposable jar with lid
- 2 disposable plastic spoons
- Lye (*sodium hydroxide*, available at hardware stores)
- Copper-sulfate solution (available at chemical suppliers)
- Food samples: egg white, milk, hamburger meat, bread, canned kidney beans, canned peas, banana
- Masking tape
- Marking pen

Proteins supply the building materials needed for repair and growth of the body. Meats, eggs, milk, cheese, grains, fruits, vegetables, and many plants contain different forms of proteins. You can easily test for proteins in foods with this experiment, but the procedure involves making a lye solution (a strong base that can irritate your skin and eyes), so you'll need the help of an adult. You'll also want to use materials you can throw away when the experiment is finished.

Procedure

1. In a small bowl, separate the white from the egg (freeze the yolk to save it for baking). Place ¼ cup (60 ml) of egg white in the first plastic glass and label it "egg white." Wash the bowl.

2. Place ¼ cup (60 ml) of milk in the second plastic glass and label it.

3. Place ¼ cup (60 ml) of banana in the small bowl, add ¼ cup (60 ml) of water, and mash the banana-water mixture until you have a paste.

4. Pour this paste into the saucepan, and heat it until it starts to boil. Add the boiled paste to the third plastic glass and label it "banana." Wash the bowl.

5. Repeat steps 3 and 4 for the kidney beans, peas, bread, and hamburger. Label all the drinking glasses clearly.

6. You MUST have an adult's help for this step. Pour ½ cup (120 ml) of warm water into the jar. Use the disposable plastic teaspoon to add lye until no more lye dissolves in the water. Put the lid on the jar and label the jar: "LYE SOLUTION: CAUTION." Discard the plastic spoon.

Caution: REMEMBER, LYE CAN IRRITATE YOUR SKIN AND MUST NOT BE BREATHED IN. KEEP YOUR FACE WELL AWAY FROM THE POWDERED LYE AS YOU SPOON IT INTO THE WATER.

7. Move the lye solution close to your row of food samples. Carefully remove the lid from the jar, and use the eyedropper to transfer lye solution into each of the plastic glasses containing food. Each glass should contain about as much lye solution as food. Rinse the eyedropper thoroughly before reusing.

8. Use the second disposable plastic teaspoon to mix the food samples in each glass. Make sure that you rinse off the teaspoon between mixings.

9. Use the eyedropper to add 4 drops of copper-sulfate solution to each of the glasses, and observe any changes in color.

Note: If you cannot find copper sulfate in solution form, mix 5 copper-sulfate crystals in ½ cup (120 ml) of water. Although you should always practice caution when handling chemicals, copper sulfate is not a harmful substance.

Caution: AS A SAFETY PRECAUTION, WHEN YOU HAVE FINISHED YOUR EXPERIMENT, DISPOSE OF THE EYEDROPPER, JAR, GLASSES, AND BOWL.

Result: Some of the food samples turn a brilliant violet color, while others are less violet or show no color change at all.

Explanation

The food samples containing the most protein turned a brilliant violet color when you added the copper-sulfate solution. Foods with less protein show less color. The banana showed no color change due to the small amounts of protein in that fruit. Boiling solid foods breaks down the protein so that you can more easily test for it.

Display Tip

Use photographs to document each stage of your experiment, displaying them (with captions) in your booth. Use plain water and purple food coloring to reproduce the color changes associated with high and low quantities of protein. Place a photograph or drawing of each type of food you tested before its appropriate color.

Did You Know?

Although many vegetables, legumes, grains, and dairy products contain proteins, most of these are incomplete proteins. This means that the body needs to combine one type of incomplete protein with another type in order to get the same amount of nutrition you would obtain by, say, eating a piece of meat or fish.

You can get complete proteins from combining certain foods, and many cultures throughout the world have found ways to satisfy all their protein requirements from a mostly vegetarian diet. A common delicious dish mixes rice with beans. The incomplete proteins in the rice (grain) and the beans (legume) combine to make a complete protein, and the tasty result is about as protein-rich as a ham-and-egg omelet. Also, there's a reason cereal tastes better with milk. The incomplete proteins of both the grain and dairy combine to create an energy-rich, protein-packed breakfast.

Why Do We Cook Food?

You Will Need
• Raw chicken leg
• Raw lamb chop
• 2 forks
• Teaspoon (5-ml spoon)
• Cooking pot
• Pot holder
• Plate
• Rubber gloves
• Adult help

Who was the first person to place a piece of raw meat over a fire? Who was the first to boil a vegetable? An ancient Chinese story tells of a farmer whose barn burned down. After the fire, the farmer entered the barn and smelled something good. He brought cooked meat back to his family that evening. In reality, the practice of cooking meat probably goes back many thousands of years, long before people built barns!

Cooking causes chemical changes that make many foods easier, tastier, and safer to eat. Heat causes some proteins in raw foods to harden, or *coagulate*. This is what happens when you cook an egg. The albumin in the egg becomes solid. And this is also why cooks use eggs to "hold up" their cakes and pudding desserts. Heat causes other proteins to break down so that the food becomes softer. When proteins break down, they create other chemicals that give cooked foods their special flavors.

By cooking two small pieces of meat, you can learn more about how heat changes proteins.

Procedure

1. Begin by examining the raw chicken leg. Put on the rubber gloves since raw chicken can contain bacteria (salmonella and more) that may harm you. Clean up with hot soapy water.

2. Place the chicken leg on the plate, and try to pull the leg apart with the two forks.

3. Put the chicken leg in a pot, and fill the pot with just enough water to cover it.

4. Boil the chicken leg for about 30 minutes.

5. Remove the chicken leg from the pot, and place it on a clean plate. Pull the leg apart with the two forks, and examine the bones and meat. Wash out the pot, and use the cooked chicken for chicken salad.

6. Place a raw lamb chop on the plate, and try to separate the meat from the bone with just the forks.

7. Put the lamb chop in the pot. Fill the pot with just enough water to cover it.

8. Boil the lamb chop until most of the water has evaporated.

9. Remove the lamb chop from the pot and place it on the plate. Use the forks to pull the meat apart, and separate it from the bone. You can mix the well-cooked meat in an omelet or use it for a sandwich later.

10. Allow the water in the pot to cool; then use the teaspoon to skim off the white layer of fat.

11. Examine the jellylike stuff left in the pot.

Result: After boiling, both the chicken leg and lamb chop easily fall apart and separate from their bones. After it cools, the water you boiled the lamb chop in forms a layer of fat over a jellylike substance.

Explanation
Raw meat is held together by connecting tissues made of proteins. Bones are held together by special proteins called *ligaments*. Cooking dissolves both of these proteins. The raw chicken leg was attached firmly to the bone and almost impossible to pull apart. After boiling the leg for 20 minutes, the meat could be pulled away and the bones separated.

The tough lamb chop also grew softer after it boiled, eventually falling apart. As you continued to boil the meat, more and more protein dissolved until there was little left to hold the meat together. But in the case of the lamb chop, the dissolved protein turned into a jellylike form that you could only see when the water cooled. This new form of protein, *gelatin*, is made entirely of dissolved tissue protein.

Display Tip
Record your cooking experiment with photographs, listing all the changes you observed as raw food cooked.

The Mighty Starch Molecule

You Will Need

- 2 raw potatoes
- Potato grater
- Cloth napkin
- Bowl of water
- Soup ladle
- Saucer
- Tablespoon (15 ml spoon)
- Coffee filter
- 4 small drinking glasses
- Measuring cup
- Saucepan

Scientists believe that the starch molecule is one of the biggest and most complicated food molecules that exists in nature. A microscope shows that grains of raw starch are large and that they have tough walls to keep them from soaking up cold water. Boiling softens and bursts the walls so that the grains can combine with water. This is why a cooked potato is so much easier to eat than a raw one.

When starch combines with hot water, the starch molecule breaks up into a smaller molecule called *dextrin*—a kind of sugar. If you then heat dextrin, it breaks up into an even smaller sugar molecule called *maltose*.

In this project, you'll first make starch and then turn it into a sugar by heating.

You can combine this project with Iodine Test for Starch.

Part 1 Homemade Starch

Procedure

1. Grate the raw potatoes into the cloth napkin.
2. Twist the napkin into a ball around the grated potatoes, and hold the napkin under water.
3. Squeeze the ball for a few minutes so that the cloudy starch comes through the napkin.
4. Remove the ball and throw away the potato gratings.
5. Half-fill glass #1 with the starchy water and put it aside for now.
6. Place a coffee filter in glass #2.
7. Using the soup ladle, carefully pour the starchy water through the coffee filter until glass #2 is filled. Then discard the filtered water from the glass, and continue ladling starchy water. A layer of sticky potato starch will stick to the sides of the filter.
8. Allow the filter to dry overnight or until the starch becomes a thin white crust.
9. Remove the crust from the coffee filter, break it into small pieces, and place the pieces in a saucer.

Part 2 Heating Starch

Procedure
1. Put 2 tablespoons (30 ml) of starch in the saucepan, and gradually heat it until it turns brown.
2. Add 1 cup (240 ml) of water and stir.
3. Half-fill glass #3 with this liquid, and pour the remaining mixture through a coffee filter into glass #4.
4. Allow the filter to dry.
5. Taste the substance left on the filter. Does it remind you of something?

Explanation
Starch is an energy-giving food. Most plants store their energy in the form of starch, and many parts of plants, like roots and seeds, are rich in starch. Scientists believe that a starch molecule is much larger and more complicated than a sugar molecule, even though starch and sugar are very similar chemically.

When you heat a starch molecule, it breaks apart into smaller sugar molecules that can be more easily digested. In plants, this process occurs with the help of the Sun's energy. *Photosynthesis* helps a plant break down stored starch into the sugars the plant needs to survive.

If the starch left on the coffee filter tasted familiar, that's because heating turned it into the sugar dextrose, used as a glue to seal envelopes.

Display Tip
Document your starch experiments with photographs. List starch-rich vegetables and describe what happens when you cook them. Do you think corn and rice would be very digestible grains before cooking?

Did You Know?
Many instant foods, such as instant potatoes and 1-minute rice, have already been cooked to break apart the mighty starch molecule.

Iodine Test for Starch

You Will Need

- 2 raw potatoes
- Potato starch water (see The Mighty Starch Molecule project)
- Soup ladle
- Coffee filter
- 2 small drinking glasses
- Small cooking pot
- Iodine *(see caution, p. 90)*

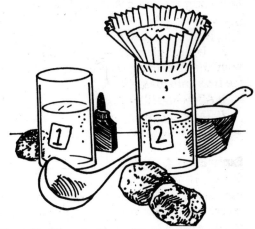

You can combine this project with The Mighty Starch Molecule.

Procedure

1. Half-fill glass #1 with starch water, and put it aside for now.
2. Place a filter in glass #2.
3. Half-fill glass #2 with this liquid.
4. Using the soup ladle, carefully pour the starchy water through the coffee filter until glass #2 is filled. Then discard the filtered water from the glass, and continue ladling starchy water. A layer of sticky potato starch will stick to the sides of the filter.
5. Allow the filter to dry overnight or until the starch becomes a thin white crust.
6. Remove the crust from the coffee filter, break it into small pieces, and place the pieces in a saucer.
7. Put 2 tablespoons (30 ml) of starch in the pot, and gradually heat it until it turns brown.
8. Add 1 cup (240 ml) of water and stir.
9. Half-fill glass #1 with this liquid, and pour the remaining mixture through a coffee filter into glass #2.
10. Add 3 drops of iodine to each glass, and watch for changes.

Result: The iodine turns the water in glass #1 deep blue, and it turns the water in glass #2 deep red.

Explanation

Glass #2 contains the cooked starch, which has turned into *dextrin,* a sugar. Dextrin still reacts with iodine. Starch is an energy-giving food. Most plants store their energy in the form of starch, and so, many parts of plants, like roots and seeds, are rich in starch.

When you heat a starch molecule, it breaks apart into smaller sugar molecules that can be more easily digested. In plants, this process occurs with the help of the Sun's energy. *Photosynthesis* helps a plant break down stored starch into the sugars the plant needs to survive.

Taste the substance left on the coffee filter. Does it remind you of something? Envelopes and stamps both use dextrin for glue. Licking a postage stamp means ingesting about 8 calories.

Display Tip

Show off the results of your iodine tests. List starch-rich vegetables and describe what happens when you cook them. Do you think corn and rice would be a very digestible before cooking?

Why Toast Tastes Better

You Will Need

- Slice of white bread
- Bowl of water
- Toaster
- Iodine
- Aluminum foil
- Plastic knife

Adding heat causes many of the chemicals in foods to change. Some of these changes make the food softer and easier to digest while others result in pleasant tastes. In this project, you can see how simple toasting leads to chemical changes in bread that make for a tastier breakfast.

Procedure

1. Mix about 5 drops of iodine into a small bowl of water.
2. Toast the slice of bread.
3. Cut the slice into strips, and quickly dip one of the strips into the bowl. Remove the strip and place it on the aluminum foil.
4. Dip a second strip into the bowl, remove it, and put it on the aluminum foil.
5. Compare the two strips of bread.

Result: In both cases, the white untoasted edge of the bread turned bluish-purple from the iodine, indicating starch. The toasted surface either didn't change color, or turned slightly red to indicate the presence of dextrin.

Explanation

Toast tastes better because toasting turns the starch at the surface of the bread to dextrin or sugar. Dextrin has a pleasant and mildly sweet taste. The crusts of baked goods also contain dextrin, formed from starch during the baking.

Display Tip

Exhibit your colorful bread samples along with an explanation of what happened. Test other baked goods for dextrin and list your results.

Iron Content in Fruits & Vegetables

You Will Need

- 1-pint (500-ml) jar
- 2 black tea bags
- Fresh asparagus
- Fresh spinach
- Fresh (if possible) pineapple juice
- Apple juice
- White grape juice
- Cranberry juice
- 6 clear plastic cups
- Measuring spoons
- Warm water
- Blender
- Coffee filter

In order for our bodies to remain healthy, we need small amounts of the mineral iron so that our red blood cells can produce *hemoglobin*. This substance, in turn, allows the red blood cells to carry oxygen to other cells in the body. Many foods contain iron, particularly green leafy vegetables, red meats, and some fruits. You'll test some of these foods in this project.

You can combine this project with Testing for Iron in Breakfast Cereals.

Procedure

1. Make a strong tea solution by placing the two tea bags in a jar filled with warm water and leaving it undisturbed for at least 1 hour.
2. Have an adult help you make a juice from the asparagus and spinach. Add 3 tablespoons (45 ml) of water to the blender, then cut up 1 cup (240 ml) of each vegetable, and blend them separately. Strain each juice through a coffee filter to remove particles.
3. Pour the asparagus and spinach juice into separate plastic cups; pour the pineapple, apple, grape, and cranberry juice into separate cups, also.
4. Add 4 tablespoons (60 ml) of tea to each cup, rinsing the spoon clean after each use.
5. Leave the cups undisturbed, but observe each cup after 20 minutes, making note of any changes in the cups.
6. After 3 hours, reexamine each cup, and look for dark particles that have settled at the bottom.

Result: Within 20 minutes, dark particles form in the asparagus, spinach, and pineapple juice. Dark particles form in the cranberry and grape juices after 3 hours. No particles form in the apple juice.

Explanation

Adding tea to the juice samples caused a chemical change that indicates the presence of iron. The tannic acid in tea combines with the iron to produce dark particles. Particles form quickly in asparagus, spinach, and pineapple juices because these foods contain high concentrations of iron. Particles form more slowly in cranberry and grape juices because these juices contain less iron. Apple juice contains no iron and does not react with the tea solution.

Display Tip

Document your experimental procedure with photographs. Also reconstruct your experiment in your exhibit and display results of your iron tests in a chart.

Did You Know?

Scientists have long wondered how the magnetic form of iron called *magnetite* gets deep inside rocks. Now they have discovered the answer in the form of a strange bacteria that seems to have quite an unusual lifestyle. It breathes without oxygen, enjoys the dark, and loves to eat iron. When billions of these tiny organisms digest iron, they turn it into magnetite. When the organisms die, the magnetite forms into layers within the rock.

Iron Content in Breakfast Cereals

You Will Need

- 3 samples of breakfast cereal; one "iron fortified" or with 90% U.S. RDA (recommended daily allowance) for iron
- 3 plastic sandwich bags
- 3 saucers
- Bar magnet
- Coffee filter

Procedure

1. Place three samples of breakfast cereals in three plastic sandwich bags.
2. Use your hands to crush each cereal into a fine powder.
3. Place the three powders in separate saucers, labeling the iron-fortified sample.
4. Dip one end of the bar magnet in the first two samples, the unfortified ones. Remove the magnet and examine it.
5. Dip one end of the magnet in the fortified cereal. Remove it and examine it.

Some solid foods have tiny quantities of iron added to them. But without a chemical indicator like the tea in the fruit and vegetable juice test, it's very difficult to isolate the added iron. Cereal is the exception, as you'll see in this project.

You can combine this project with Iron Content in Fruits & Vegetables.

Result: When you remove the magnet from the iron-fortified cereal, tiny white particles of cereal containing *edible iron* cling to the magnet. This doesn't happen in the unfortified cereals. Crushing the cereal into a fine powder allows the traces of iron to drop away from the rest of the solid-food particles and to stick to the magnet.

Bones & Minerals

> **You Will Need**
>
> - 2 mayonnaise jars (or similar), same size
> - 2 cooked chicken bones, small enough to fit inside the jars
> - White vinegar
> - Tape and marker for labeling

We know that proteins, carbohydrates, and fats all combine to keep you healthy. We also know that vitamins are important for cell growth and cell repair. But what about minerals? What would happen if your body didn't receive the proper amounts of minerals, or no minerals at all? This project helps you find out.

Procedure

1. Fill one jar three-fourths full of vinegar and the other three-fourths full with water. Label the jars with the tape and marker.
2. Wash the chicken bones with soap and water and allow them to dry.
3. Place one of the bones in the vinegar jar and the other in the water jar.
4. Wait 1 week and remove the bones from the jars. Allow them to dry.
5. Try to bend the bone that you removed from the water; then try to bend the bone you removed from the vinegar.

Result: The vinegar-soaked bone easily bends as if it were made of rubber. The water-soaked bone is unchanged.

Explanation

Vinegar, an acid, dissolved most of the mineral calcium. Without this mineral support, the vinegar-soaked bone became weak, rubbery, and useless. Water had no effect on the calcium in the other bone. This doesn't mean that you should leave out the vinegar the next time you have a salad, but that you should always drink your milk.

Display Tip

Your rubberized bones make an interesting and amusing display. Use photographs to document how you performed this experiment.

Did You Know?

Some minerals have other functions besides building bones. The mineral iron, for example, helps your tissue stay healthy and provides the energy you need to stay active. Iron helps your body make hemoglobin, or the stuff in red blood cells. Hemoglobin helps carry oxygen to your cells. In turn, your cells use the oxygen to produce energy. How do you think you would feel without enough iron in your diet?

Excellent Electrics

A Springing Spring
Copperplating a Key
Electroscope
How to Read an Electric Meter
Jumping Puffed Wheat
Removing Silver Tarnish
Flash Dancers
Induction Coil & Galvanometer
Controlling Current
Electrostatic Flower
Strange Stocking Trick
Eddy-Current Motor
Potato Polarity Indicator

A Springing Spring

You Will Need

- Thin copper wire
- Thick copper wire or copper nail
- 2 pieces of insulated wire
- Small Styrofoam ball
- Small bowl
- Sharp pencil
- Salt
- 6-volt battery
- Stack of books
- Masking tape
- Scissors

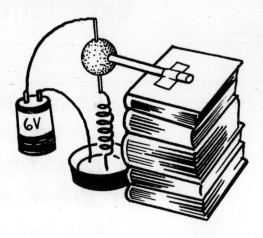

By passing an electrical current through a copper wire, you can make the wire behave in mysterious ways. This project demonstrates how a coil of wire behaves like a magnet when carrying an electrical current through a simple closed circuit.

Procedure

1. Twist the thin copper wire around the pencil in a coil, but avoid overlapping the wire. Carefully slip the coil off the pencil.
2. Carefully push the thick copper wire through the Styrofoam ball so that the wire sticks out at both ends. Push the sharp end of the pencil into the side of the cork.
3. Tape the other end of the pencil to the side of a book. Place this book at the top of a pile of books so that the pencil, Styrofoam ball, and wire hang out from the stack of books by about 5 inches (12.5 cm).
4. Gently stretch out the coil of copper wire so that it becomes a kind of spring. Straighten out each end of the spring, and twist one of the ends around the thick copper wire sticking out from the bottom of the Styrofoam ball.
5. Have an adult help you strip the plastic insulation from the ends of both pieces of insulated wire. Attach each piece of wire to a terminal of the battery.
6. Attach the opposite end of one of the wires to the thick copper wire sticking out from the top of the Styrofoam ball.
7. Fill the small bowl with warm water, and add salt until no more salt dissolves. Place the end of the copper spring in the saltwater so that the wire sits just below the water's surface.
8. Take the disconnected piece of insulated wire from the battery, and gently dip the stripped end into the saltwater.

Result: When the battery-connected wire touches the saltwater, the copper spring bounces wildly. The bouncing continues until you remove the wire from the water.

Explanation

The battery, wires, copper spring, and saltwater all create a simple circuit. When you touch the end of the battery-connected wire to the saltwater, you complete the circuit so that electricity travels in a continuous loop. Electrical current passing through the spring turns it into a magnet with a north and south pole.

A Springing Spring

Since north and south poles attract each other, the spring pulls together. But when this happens, the end of the spring pops out of the water, breaking the circuit and stopping the flow of electricity. With no electricity, the spring no longer behaves like a magnet and stretches out so that the end once again dips into the saltwater. This restores the electrical circuit so that the spring becomes a magnet again and the bouncing continues.

Display Tip

This is an excellent working model to exhibit for the judges, and it will entertain everyone who visits your booth. Document the construction of your springing spring with photographs, and mount the photographs behind the model. Provide a clear explanation of the principles of *electromagnetism* at work.

Did You Know?

Electrolyte means any liquid that conducts electricity. Acids, such as vinegar, lemon, or tomato juice, make excellent electrolytes. So do many salts, including neutral salt (sodium chloride or table salt), acid salt (cream of tartar), and basic salt (baking soda).

Copperplating a Key

You Will Need

- Brass key
- Bendable copper strip
- Cardboard milk container
- Scissors
- 6-volt battery
- Insulated copper wire, cut into equal lengths
- White glue
- Cellophane tape
- Paper clip
- Vinegar
- Saltwater

The technique of electroplating demonstrates important principles of conductivity. It also turns a dull key into a brilliant copper beauty.

Purpose
To show the interaction between active metals in an electrolyte bath.

Procedure

1. Use the scissors to cut the milk carton down into a tray about 4 inches (10 cm) deep.
2. Fill the tray with just enough vinegar to cover the key as if it were standing on end. Add salt to the vinegar until no more salt dissolves.
3. Have an adult strip the ends of the insulated copper wire. Tape one end of one wire to the positive terminal of the 6-volt battery. Tape the other end of that wire to the strip of copper.
4. Bend the copper strip so that it clips onto the side of the milk container with at least 3 inches (7.5 cm) of it hanging into the vinegar-salt solution.

Electroplating Setup

5. Wash the brass key with a little dishwashing detergent to remove dirt and grease.
6. Loop one end of the second piece of wire through the hole in the key, twisting it securely. Connect the other end or the wire to the negative terminal of the battery.
7. Bend the paper clip into an L shape, and tape one end to the side of the milk carton so that the other end hangs out over the vinegar-and-salt solution like a fishing rod.
8. Hang the key end of the wire from the fishing rod so that the key is submerged in the solution. To keep the wire from slipping, loop it once or twice around the paper clip.
9. Observe any changes in the key over the next hour.
10. Remove the key and copper strip from the solution and examine them.

Result: In about 20 minutes, the key begins to turn a slight coppery color. In an hour, it is thoroughly copperplated. When the copper strip is removed from the solution, it feels brittle and crumbly.

Explanation
As an electrical current passes through the vinegar-and-salt solution and into the metals, one metal loses its electrons

to the other metal. Scientists call metals that lose electrons *active metals*. In this case, the more active copper lost electrons to the less active brass. The copper strip became brittle because much of it was literally used up by the brass. If the copperplating had continued for several hours, the copper strip would have completely dissolved in the solution.

Display Tip
Copperplate several small brass items, and display them alongside a mock-up of your electroplating bath. Draw a diagram that clearly describes the interactions between active and less active metals in an electrolyte solution.

Did You Know?
In electroplating, the more active metal (here, copper) is called the *anode* and the less active metal (here, brass) is called the *cathode*.

You can also electroplate nickel over copper, since nickel is the more active metal of the two.

Electroscope

You Will Need

- Small glass jar
- Wire coat hanger
- Aluminum foil
- Cardboard
- Strip of silver Mylar from helium balloon or wrapping paper
- Plastic comb
- Balloon
- Pencil
- Cellophane
- Rubber cement
- Electrical tape
- Wire clippers
- Pliers

An electroscope is a simple device that measures *static electricity*, or the freely flowing electrical charges of the atmosphere. Static electricity is caused by friction, or something rubbing against something else. This electricity-producing friction can be as harmless as a comb rubbing against hair or as powerful as ice crystals rubbing against each other in a thundercloud.

Procedure

1. Use the wire clippers to cut off a straight piece of wire from the hanger. Use the pliers to bend a section at one end into an L shape.
2. Turn the jar upside down onto the cardboard, and trace a circle around the opening.
3. Cut out the circle and punch a small hole in its center with a pencil.
4. Carefully push the wire through the hole about 1 inch (2.5 cm), straight end first.
5. With as little cellophane tape as possible, attach the middle of the Mylar strip to the bent end of the wire so that the strip hangs down in two equal halves.
6. Rubber cement the cardboard circle to the top of the jar, with the bent end of the wire holding the Mylar strip pointing down.
7. Place electrical tape around the edges where the circle touches the rim of the jar, and place a thin band of cement around the wire where it punches through the cardboard.
8. After the rubber cement dries, crumple the aluminum foil into a tight ball, and carefully push it onto the top of the wire.
9. Rub the plastic comb or balloon against your hair or clothing (wool works best), and hold either one close to the aluminum foil ball.

Result: If the air is dry enough, the ends of the Mylar strip fly apart when the comb or balloon touches the ball.

Explanation

An electroscope shows the attraction and repulsion of electrical charges. In all electrical activity, *like charges* repel and *opposite charges* attract. When you rub the comb, friction causes a positive charge to build up in the plastic. When you hold the positively charged comb near the aluminum-foil ball, the comb attracts negative charges which move up through the wire so that only positive charges remain in the Mylar strip. Since both ends of the strip now have the same charge, the ends of the strip fly apart.

Display Tip

Display your electroscope along with an assortment of objects for creating static charges.

How to Read an Electric Meter

You Will Need

- Pad and pencil

This project will teach you how to read that mysterious device attached to the outside of your house: the electric meter. Reading the meter will help you calculate the number of *kilowatt-hours* you use and learn how to conserve energy in your home.

Your house uses electrical power 24 hours a day. When the electric company sends a bill, the amount you owe reflects the amount of electricity used in one month. The company knows what to charge by reading the meter's dials and doing some simple arithmetic.

Procedure

1. Locate the electrical meter outside of your house or apartment. Look for a small wall-mounted instrument with four dials, covered by a circular glass window.
2. Look closely at the four clockface dials at the top of the meter. The dials are numbered 0–9 and alternate in a clockwise and counterclockwise direction.
3. Read the arrow of each dial from left to right, and determine the four-digit number indicated by the set of dials. When the dial points between two numbers, take the lower number. For example, the dials on p. 119, bottom left, show the number 2,493.
4. Write down the number on your dial. This number represents the total amount of kilowatt-hours used.
5. Wait a week and read the dials again. Write down the new kilowatt-hour number.
6. Subtract the first number from the second number. The difference indicates the number of kilowatt-hours you used in a week.
7. Wait another week and read the meter again. Subtract last week's reading from your new reading. Did you use more or less power than the week before?
8. Before reading the meter one more time, think of ways to conserve electricity. Turn off lights when you don't use them, and avoid using the toaster or space-heater. If possible, do without the electric iron or electric dryer.
9. Compare your last reading with the readings of the weeks before.
10. Look at a recent electric bill to see what your electric company charges for each kilowatt-hour. (In the United States, it's usually about 10 cents.) Multiply 10 cents by the number of kilowatt-hours on each bill, and divide by 100 to see what you paid in dollars each month. Compare those totals to see what you saved. *Or, figure the cost in your country's currency.*

Result: Using fewer electrical appliances results in less power consumption and a smaller number of kilowatt-hours displayed on the meter. This means that you owe less money to the electric company.

Explanation

Each month, electric companies know what to charge you by sending someone out to read the total number of kilowatt-hours displayed on your meter. Think of a kilowatt-hour as a unit of measurement rather than a unit of time. Each kilowatt hour represents 1,000 watts of consumed electrical power. If you leave only one 25-watt lightbulb burning for 40 hours, it consumes 1,000 watts, represented as 1 kilowatt-hour on your meter. If you leave a 500-watt dryer on for 2 hours, it also uses 1,000 watts or 1 kilowatt-hour. If your electric company charges 10 cents per kilowatt-hour, running the dryer for 2 hours or burning the lightbulb for 10 hours would cost the same: 20 cents.

To consume power, however, most electric companies have a base-usage limit on kilowatt-hours. This means that your home is allowed only a certain number of kilowatt-hours per month, and that any kilowatt hours above the base-usage limit will be charged at a higher rate.

A *watt* is a very small unit of power consumption. On an electric meter, the dial at the far right moves around quickly because it measures single watts; the dial next to it measures watts in groups of 10; the next dial measures watts in groups of 100; and one complete revolution of the leftmost dial indicates 1,000 watts, or 1 *kilowatt.* The kilowatt dial is the first dial the meter reader looks at to calculate your total number of watt-hours.

Each complete turn of a dial causes the next dial to turn, and the gear mechanism between neighboring dials is the reason they alternate in clockwise and counterclockwise directions.

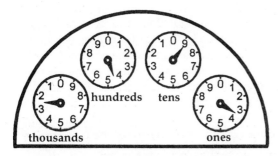

Meter Reading: 2,493 kilowatt-hours

Display Tip
Make a weekly chart showing the kind and number of devices you avoided using and the number of watt-hours you saved. Calculate how much electrical power could be saved over six months if ten households practiced conservation. Do you think we could use this conserved power in better ways?

Did You Know?
Consult the chart below for the power, estimated in watt-hours, of common household electrical appliances.

ELECTRICITY CONSUMPTION OF SMALL HOUSEHOLD APPLIANCES

Watts Consumed Per Hour

Air-Conditioner (room)	1,389
Blender	15
Broiler	100
Clock	17
Clothes Dryer	993
Clothes Washing Machine	103
Coffee-maker	106
Computer	25–400
Dehumidifier	377
Dishwasher	165–363
Fan (circulating)	43
Fan (attic)	291
Food Mixer	13
Freezer (frost-free)	1,820
Frying Pan	186
Garbage Disposal	30
Hair Dryer	14
Iron	144
Microwave Oven	300
Radio	86
Refrigerator-Freezer (frost-free)	1,591–1,829
Stove (with self-cleaning oven)	1,205
Television (black-and-white)	362
Television (color)	502
Toaster	39
Vacuum Cleaner	46
Video Cassette Recorder (VCR)	10–70
Water Heater (standard)	4,219

Jumping Puffed Wheat

> **You Will Need**
> - A few kernels of puffed-wheat cereal
> - Saucer
> - Plastic comb
> - Plastic toothbrush box
> - Piece of wool cloth

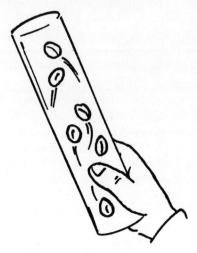

This project shows the attraction and repulsion of charged electrons by producing static electricity in puffed wheat.

Part 1 Combing Electrons

Procedure
1. Place a few kernels of puffed-wheat cereal into a saucer.
2. Rub the plastic comb for a few seconds with the piece of wool cloth.
3. Drag the comb across the puffed wheat in the saucer. Several kernels will stick to the comb.
4. Gently pull the comb and puffed wheat away from the saucer. Hold the comb in front of you, and watch what happens to the kernels of puffed wheat.

Result: After about 30 seconds, the kernels of puffed wheat start popping off the comb.

Part 2 Electron Boxing

Procedure
1. Place a few kernels in the plastic toothbrush box.
2. Rub the box for about 20 seconds with the wool cloth.
3. Watch what happens to the kernels in the plastic box.

Result: The kernels stick to the sides of the box. But after about 30 seconds, they will pop off the sides of the box and continually jump around.

Explanation
Under normal circumstances, an object has the same number of negatively charged electrons and positively charged protons. This means that the object is electrically *neutral*. However, electrons and protons differ in that electrons can move about freely while protons remain stationary.

When you rub an object like a plastic comb with a wool cloth, the friction rubs electrons off the cloth and onto the comb. This means that the comb has many more electrons than the cloth and becomes negatively charged. When you held the negatively charged comb against the neutral kernels of puffed wheat, the protons in the kernels were attracted to the comb. But after a while, the negative electrons in the comb passed into the kernels so that eventually, both comb and kernels had negative charges. Since like charges repel each other, the kernels sprung away from the comb.

The toothbrush box demonstrated the same principle. Rubbing the box with the wool gave the box a negative charge which attracted the neutral kernels to the sides of the box. Eventually, the kernels took on the negative charge and were repelled from the box and from each *other*, as well. Caught within the contained space of the box, the kernels jumped around until they lost their charges.

Display Tip
Think of other ways you can demonstrate static electricity with everyday items. For example, try rubbing wool cloth against a balloon and then "sticking" the balloon against a surface. Can you think of examples of static electricity in nature?

Removing Silver Tarnish

You Will Need

- A few pieces of tarnished silverware
- 2 iron or stainless-steel pots (do not use aluminum)
- Hot plate
- Tap water
- Dishwashing detergent
- Aluminum foil
- Baking soda
- Plastic or stainless-steel tongs

Metals are chemically active substances. By chemically active, scientists mean that parts of metal atoms—called electrons—break away from the atoms and combine with materials outside the metal. Certain metals are more chemically active than others because they lose more electrons. A chemically active metal changes easily, as indicated by both the rusting of iron and the tarnishing of silver.

Since all metals are chemically active to one degree or another, you can reverse the process of electron loss and actually *add* electrons to a metal. Scientists do this by pulling the electrons from a more active metal (like copper) and "gluing" them to a less active metal (like zinc). This is the basis of electroplating.

Electroplating can also remove tarnish by replacing the lost electrons in silver. Tarnish on silver shows where the electron loss is particularly severe.

Procedure

1. Wash the silver with dishwashing detergent to remove any dirt or grease that might interfere with the plating process.
2. Line the sides and bottom of the pot with sheets of aluminum foil.
3. Place the silver objects in the pot, and fill the pot with just enough water to cover them. Remove the objects.
4. Place the pot on the hot plate, add 3 tablespoons (45 ml) of baking soda, and wait for the water to boil.
5. Grasp the silver objects with the tongs, and gently lower them into the boiling water. After a few minutes, the dark tarnish spots begin to disappear.
6. When the objects look completely silver again, carefully remove them with the tongs, and dip them into a second pot of cold water to remove any remains of baking soda.
7. Place the objects somewhere safe to cool off before handling them.
8. Use the tongs to carefully remove a piece of aluminum foil from the bottom of the pan. Rinse the foil to cool it, and then examine the foil carefully.

Result: The silver objects lose all traces of tarnish. The aluminum foil looks dull and crumbles easily.

Explanation

When you added baking soda to water, you created an *electrolyte*—or liquid that conducts electricity—through which electrons could pass freely between the aluminum foil and silver. Boiling the water created heat and speeded up the electrical interaction between the metals.

Since a more active metal loses electrons to a less active metal, the more active aluminum lost electrons to the less active silver. The aluminum foil became brittle because much of it was literally used up by the silver. Do you understand why an aluminum pot was not recommended for this project?

Display Tip

Take before-and-after photographs of your silver objects, and display them next to a nonworking reproduction of your experiment. Look around your environment for other examples of active metals reacting with their surroundings. Silver is not the only metal affected by exposure to air or moisture. Look for brass and copper objects; examine the green coating, or *patina*, on bronze objects.

Collect your objects and exhibit them, and don't forget to include that most familiar example of a changing metal, rusty iron.

Flash Dancers

You Will Need

- Wrapping tissue
- Aluminum pie tin
- Cellophane
- Balloon
- Wool yarn
- Ruler
- Scissors

Static electricity does strange things, from bending water to flashing bolts of lightning across the sky. You can also see static electricity at work in the pirouettes of paper cutouts.

Procedure

1. Measure the depth of the pie tin, and cut figures from the tissue paper slightly shorter than the depth of the tin.
2. Place the figures in the tin, spacing them evenly.
3. Stretch cellophane across the top of the tin. The tighter and more drumlike the cellophane, the better your results will be.
4. Blow up the balloon. Wad the wool yarn into a ball, and rub it against the balloon until you hear the crackling sound of static electricity.
5. Gently rub the wool across the cellophane.

Result: The figures spring upright and dance crazily. Turn off the light to *really* see your figures flash dance.

Explanation

Static electricity consists of charged particles, either mutually attracted or repelled, passing from one material to another. The friction of the wool against the balloon gave the wool a positive charge and the balloon a negative charge so that the balloon and wool attracted each other. When you moved the positively charged wool to the pie tin and rubbed it across the cellophane, the positively charged wool attracted the negatively charged cellophane. In turn, the cellophane attracted the positively charged tissue-paper figures underneath. The attraction of opposite charges made the figures spring upright and whirl around.

Induction Coil & Galvanometer

You Will Need

- Strong bar magnet
- 5 feet (1.5 m) of enameled (not plastic-insulated) copper wire
- Compass
- Drinking glass
- 4 twist fasteners
- Ruler
- Scissors

This project demonstrates one of the most important discoveries of the 19th century: electrical induction. The term *induction* means that you can create electricity in a *conductor*—any material capable of conducting electricity—by moving a magnet against it. Although it was known for some time that sending an electrical current through a conductor created *magnetism*, the fact that the opposite was also true was discovered accidentally by an English scientist named Michael Faraday. The discovery of induction meant that, by pushing magnets against coils of copper wire, electricity could be created in special machines called generators.

Procedure

1. Wrap the copper wire around a drinking glass, leaving about 18 inches (45 cm) of wire at the beginning and end. You should wind up with a thick coil of wire around the glass.

2. Slide the coil from the glass and twist four fasteners around it. You want the coil thick, firmly bunched, and compact.

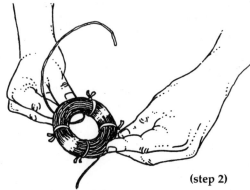

(step 2)

3. To show the flow of electricity through the coil, you have to change the compass into something called a *galvanometer*. A galvanometer shows electrical current by making the compass needle move.

4. Wrap the free ends of wire around the compass in the same direction, connecting the wires.

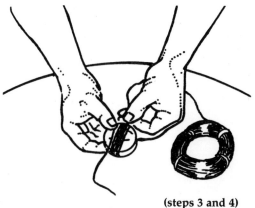

(steps 3 and 4)

5. With everything prepared, lift the coil with one hand and slowly move the bar magnet in and out of the coil's center.

Result: The compass needle jumps each time you move the magnet into the coil.

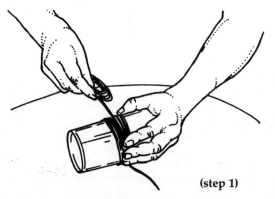

(step 1)

Induction Coil & Galvanometer

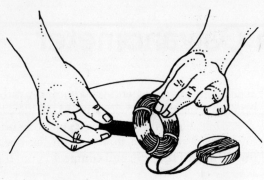

Testing the Induction Coil and Galvanometer with a Magnet (step 5)

Explanation
Each time the magnet passes the coil, the magnetic field surrounding the magnet creates an electrical current in the wire—*induction*. Each time you remove the magnet, the current stops. When you move the magnet in and out of the coil quickly, it creates a pulsing stream of electrical current called *alternating current*. This is the type of electricity most commonly used today.

Display Tip
Record the construction of your induction coil and galvanometer with photographs. Exhibit your working model alongside pictures of modern generators. Research Michael Faraday's life, and list some his important contributions to modern science.

Controlling Current

You Will Need

- Very soft (#1 or #2) pencil
- 6-volt battery
- 6-volt flashlight bulb (screw base)
- 6 feet (2 m) of insulated copper wire
- 2 paper clips
- 3 thumbtacks
- Duct tape
- Two 2 × 6-inch (5 × 15-cm) pieces of wood ½-inch (1.25-cm) thick

You might have in your home a special light switch that allows you to dim a light as well as shut it off. This switch, called a *rheostat*, works by passing electrical current through a poor conductor. The amount of current that passes through a rheostat circuit depends on how much of this material you place between the two points of electrical contact. This project demonstrates this principle through the construction of a working rheostat.

Part 1 Making the Lamp Tester

Procedure

1. Straighten out the two paper clips, and at one end of each, form a loop somewhat smaller that the diameter of the bulb's base.
2. At the other end of each paper clip, make a tiny loop that will go around the pin of the thumbtack.
3. If your thumbtacks are painted, scrape all the paint from the top of the third thumbtack.
4. Cut two 1-foot (30-cm) sections from the 6 feet (2 m) of copper wire, and scrape 2 inches (5 cm) of insulation from the ends of these sections.
5. Wrap four turns of the cleaned-off ends of one of these wires around the scraped thumbtack. Press this thumbtack into the center of the piece of wood.
6. Arrange the other two paper clips so that both large loops are exactly above the scraped thumbtack in the middle of the piece of wood. Attach the clips to the wood with the other two thumbtacks, following the illustration below.

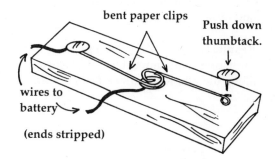

Lamp Tester (part 1, steps 5–7)

7. Around one of these tacks, wrap the cleaned-off end of the other wire to make contact with the thumbtack, and in turn, the paper clip.
8. Screw the bulb into the turned-up loops of the paper clips at the center of the wood. Make sure the bottom of the bulb touches the scraped thumbtack, adjusting the paper clips if necessary.

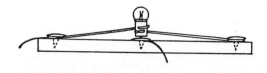

Finished Lamp Tester (*side view with flashlight bulb in place*) (part 1, step 8)

Part 2 Constructing the Rheostat

Procedure

1. Have an adult help you split the pencil so that you expose the graphite interior. Make sure you don't break the graphite.
2. Attach the pencil to the second piece of wood with two small strips of duct tape at each end.
3. Cut the remaining 4 feet (1.2 m) of copper wire into three roughly equal sections. Strip the insulation from the ends of all pieces.
4. Attach the wires to the battery, lamp tester, and pencil according to the illustration below. Leave one end of wire free.
5. Move the unattached piece of wire along the length of graphite in the pencil and observe the bulb.

Result: The light of the bulb grows stronger the closer you move the unattached wire along the graphite to the attached wire.

Explanation

Since the graphite is a poor conductor of electricity, the more of it you place between the two contact wires, the less current will flow from the battery into the bulb.

Display Tip

Document the construction and testing of your rheostat with photographs. Display your finished model with an explanation of how it works. Can you think of other devices that use rheostats? Find photographs of these devices and display them. Could you find new ways to use rheostats? Make a list of your ideas.

Did You Know?

Rheostats are not only used to regulate the intensity of light. They come in handy for controlling the volume in radios, televisions, and CD players. They also come in handy for controlling the amount of heat an electric dryer produces or the speed of a toy electric car or train. Wherever you need to turn down the current, a rheostat does the trick.

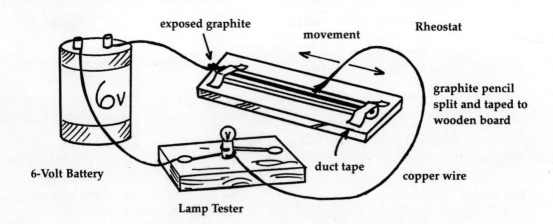

Finished Rheostat & Lamp Tester

Move wire back and forth along graphite to create a current. **(part 2, step 5)**

Electrostatic Flower

You Will Need

- 1-foot (30-cm) piece of wire hanger
- Large piece of thin tissue paper
- Balloon
- Pencil
- Duct tape
- Polyethylene plastic bag (dry-cleaning bag)

You can demonstrate static electricity in many ways. This project uses an amusing electrostatic flower model to show how like charges repel.

Procedure

1. Bend the ends of the hanger into two small loops.
2. Cut the tissue paper into eight strips, 10 inches (25 cm) long and ¼ inch (0.6 cm) wide.
3. Bundle the strips together and push them through one of the loops on the hanger so that equal lengths of tissue paper stick out from each side. Tighten the loop to keep the tissue paper from falling out.
4. With duct tape, attach the eraser side of the pencil the mid-point of the hanger. Make sure the hanger is firmly attached to the pencil and that you can lift the flower by holding its pencil "handle."
5. Blow up a balloon, and rub it with the polyethylene plastic for about 1 minute.
6. Hold the charged balloon in one hand, and, holding the flower by its pencil handle, touch the empty-hook side against the balloon.

Result: The "petals" of your flower stick out in all directions. When you break contact between the balloon and the hanger, the petals become limp again.

Explanation

All substances have positively charged electrical particles called protons and negative particles called electrons. When an object is left undisturbed for a period of time, the protons and electrons balance each other so that the object has a neutral charge.

But when you rub objects with certain other objects, electrons move into either object from the other. This means that one of two objects rubbed together becomes more negatively charged than positively charged, and that, since negative and positive charges attract each other, the objects are drawn together. Since opposite charges attract, like charges repel.

Rubbing the balloon with the polyethylene gave the balloon a strong negative electrical charge. This charge is transferred through the hanger and into the petals of the flower so that each petal also becomes negatively charged. Since like charges repel, each petal stands straight and pushes away from its neighbor.

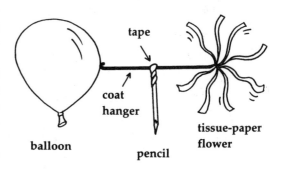

Electrostatic Flower

Display Tip

Display a working model of your electrostatic flower along with a description of the fundamental principles of static electricity that cause the flower to behave as it does.

Strange Stocking Trick

You Will Need

- Nylon stocking
- Polyethylene plastic bag (dry-cleaning bag)

This project uses a nylon stocking to show some of the strange properties of static electricity.

Procedure

1. Taking it by the toe, hold the nylon stocking against a wall, and rub it with the crumpled-up polyethylene bag. Rub with good pressure and in long downward strokes.
2. Gently move the stocking away from the wall. Try to keep it from touching any other surfaces or your clothing.
3. Observe what happens to the stocking.

Result: The walls of the stocking tube repel, and so the stocking fills out weirdly—as if there were a leg in it!

Explanation

All substances have a balance number of positively charged electrical particles called *protons* and negative particles called *electrons*. But when you rub objects with certain other objects, electrons move into either object from the other. This means that one of two objects rubbed together becomes more negatively charged than positively charged. As with magnetism, negative and positive attract while two of the same charges repel.

Rubbing the stocking with the polyethylene transferred negative charges from the balloon into the stocking. This means that the walls of the stocking were negatively charged and pushed away from each other, filling the stocking out.

Display Tip

With photographs, document how you performed this stocking experiment. Make a small display hook from a piece of hanger and suspend a charged stocking from it. The stocking will remain filled for several hours until the charge begins to leak out. Refresh your stocking by rubbing it against a wall.

Eddy-Current Motor

> **You Will Need**
>
> - Strong horseshoe magnet
> - Small aluminum-foil pie plate
> - Sewing needle
> - Cork or small piece of Styrofoam
> - 2 feet (60 cm) of heavy thread

Many familiar devices would not be possible without magnetism. Although it's easy to accept that electromagnetic motors help us vacuum floors or operate CD players, what other, more subtle role can magnets play in modern life? Is a magnet's attractive power its only useful feature, or could it be that what a magnet *doesn't* attract can be equally important? This project explores weak magnetic attraction. In doing so, you'll begin to recognize a familiar device.

Procedure

1. Push the sewing needle through the center of the cork or Styrofoam so that its point sticks up.
2. Balance the aluminum pie plate on the needle so that it is level and can easily spin.
3. Tie the horseshoe magnet to the piece of thread, and hold the suspended magnet above the pie plate as closely as possible without touching the plate.
4. Twist the string about twenty times while holding the magnet steady.
5. Let go of the string so that the magnet spins above the pie plate.
6. Observe what happens to the pie plate.

Result: The pie plate will begin to spin along with the magnet, but not quite as fast.

Explanation

Since a magnet only attracts iron, what can be happening here? Obviously, some slight magnetic force exists between the magnet and the aluminum. As the horseshoe magnet spins, it creates eddy currents above the aluminum pan. In turn, these currents produce a magnetic field above the aluminum pan's surface. The magnetic field revolves in the same direction as the spinning magnet and causes the pan to spin. The faster the magnet spins, the greater the eddy currents will be, and the greater the magnetic field above the pan will grow. The result? The pan spins faster and faster, although it will never keep up with the magnet.

Display Tip

Display your eddy-current motor along with a detailed description of its operation. List other common devices that might use this peculiar form of magnetic attraction.

Did You Know?

A car's speedometer operates on the eddy-flow principle. As the car moves, a magnet rotates at a certain speed, depending on the speed of the car. (This equation is worked out carefully by engineers.) This rotating magnet exerts a force on a disk made from some nonattractive metal, like aluminum. The disk contains the pointer of the speedometer which moves against a scale, indicating the car's speed.

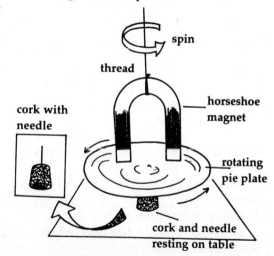

Potato Polarity Indicator

You Will Need

- D-cell battery
- 2 pieces of plastic insulated copper wire, each about 6 inches (15 cm) long
- Scissors
- Cellophane tape
- Potato
- Knife

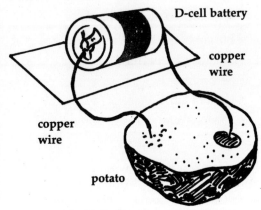

Polarity Indicator

After shopping at the grocery store, suppose your mom or dad can't get the car started and needed a battery jump. They have a jumper cable, but the battery is so old you can no longer see the positive and negative terminal indicators. Someone pulls up to help, but the owner of that car also has a very old battery. You need to figure out which are the negative and positive terminals on both batteries before you can safely connect one car to another. Can you do it? For this project, we won't use a car battery since a D-cell battery will work fine. And it's safer.

Procedure

1. Use the scissors to carefully strip away about 1½ inches (3.75 cm) of plastic insulation from each end of the wires.
2. Turn the D-cell on its side, and, with a little tape, attach one end of one of the wires to the bump side of the battery.
3. Attach one end of the second piece of wire to the flat side of the battery.
4. Cut the potato in half so that you have one wide piece. (Ask an adult's help if you aren't used to kitchen knives.)
5. Push the exposed ends of the wires into the white potato flesh, separating them by at least 3 inches (7.5 cm).
6. Observe what happens where the wires touch the potato.

Result: Small bubbles may form around one wire, and a greenish color will begin to appear around the other wire.

Explanation

The liquid in the potato reacts with electrical current in a very peculiar way. The wire attached to the positive (bump) side of the battery may cause bubbles where it comes into contact with the potato. This is because the protons of a positive charge react in a specific way to the chemicals in the potato. (This reaction isn't always very noticeable, however, and so you may not see anything.)

The wire attached to the negative (flat) side of the battery causes green to form where it touches the battery. This is a much more pronounced effect, and it results when negatively charged electrons react with the potato's chemistry. In any case, after 2 or 3 minutes, it's quite clear which is the negative end and which the positive end of your battery. On a D-cell it's easy to determine a battery's polarity, but this same simple potato test could help you find the polarity of a car battery as well.

Did You Know?

Besides a potato, the acidic juice of oranges and lemons also responds to electrical current. Scientists give the name *electrolyte* to any material that conducts the flow of current between two electrical polarities. But an electrolyte differs from a simple conductor in that it reacts more strongly to one electrical charge than to the other. This is why you may not see bubbles in the positive terminal of your potato indicator, but you'll always see green.

Natural Laws

The Strongest Bridge
Measuring Atmospheric Pressure
Foucault's Pendulum
Self-Filling Water Dish for Pets
Escape Velocity
Earth & Moon Balance
3-D Magnetic Field
Homemade Rock Tumbler
Creeping Soil
Earth Bulge
Attractive Objects
Ferromagnetic Flowchart
Magnetic Spring
The Motion of Gears

The Strongest Bridge

You Will Need

- 8 strips of shirt (thin) cardboard
- Corrugated cardboard
- 2 stacks of books, same height
- Small jar
- Bucket of sand
- Cellophane tape
- White glue
- Tablespoon (15 ml spoon)
- Ruler

It took years of trial-and-error to perfect some of the familiar bridge designs we see today. Finally, somebody had the idea that the shape of a material is even more important than its natural strength. Today engineers use very light materials to build very strong bridges.

Part 1 Sagging Bridge

Procedure

1. Place the first strip of shirt cardboard across the space between two stacks of books.
2. Test this simple *girder* bridge by placing the empty jar in the center.

Result: This simple test imitates early bridge-building experiments: a plank across water. But the cardboard bridge soon collapses from the weight of the jar.

Part 2 Archway Bridge

Procedure

1. Remove the jar, and replace the shirt cardboard strip between the stacks of books.
2. Take the second strip of shirt cardboard and bend it between the books as shown below.
3. Test this new *arch* design by placing the jar in the center.
4. Spoon sand into the jar and observe the bridge.

Result: The bridge easily supports the weight of the jar. The weight of the first strip of cardboard compresses the second piece of cardboard so that it becomes strong. As the weight increases, however, the top of the arch begins to sag.

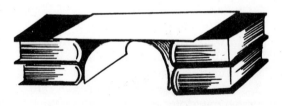

Archway Bridge (part 2, steps 1–2)

Part 3 Triangular-Truss Bridge

Procedure

1. Draw the pattern below on another strip of shirt cardboard.

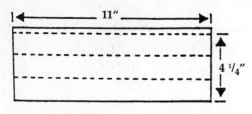

Truss Pattern (part 3, step 1)

2. Fold the cardboard into a hollow triangle and tape it together.
3. Glue this piece lengthwise to the bottom of another strip of shirt cardboard, and allow the glue to dry.

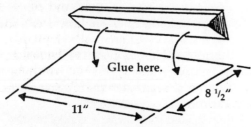

Truss Placement (part 3, step 3)

4. Test this *truss* design by filling the jar with sand as before. Observe any changes in the bridge.
5. Cut two strips from the corrugated cardboard, roughly the same size as the shirt-cardboard strips. Before you cut the strips, however, look along the edges of the corrugated cardboard. Two of the edges will have a wavy corrugation pattern. Cut your strips lengthwise from either of these edges.
6. Place a strip of shirt cardboard beside the strip of corrugated cardboard, noting the different thicknesses.
7. Using white glue, build up the shirt cardboard with additional pieces until its thickness is roughly that of the corrugated cardboard. Allow the layered shirt cardboard to dry.
8. Place the layered cardboard between the stacks of books, and place the empty jar in the center of the cardboard bridge.
9. Slowly add sand to the jar, counting out the tablespoons (15-ml spoonfuls) before the bridge collapses.
10. Replace the layered cardboard with the single strip of corrugated cardboard, positioning the empty jar as before.
11. Slowly add sand, counting the tablespoons. Observe any changes in the bridge.

Result: The truss-design bridge easily supports the weight of a full jar without showing any signs of strain. This is due to the triangular design of the large truss. The layered-cardboard bridge supports the empty jar, although it may show some bending. By making a stronger bridge by replacing the layered cardboard with one piece of corrugated cardboard, you show the additional strength of a corrugated material.

Explanation

Of all your bridge designs (girder, arch, and triangular truss), a triangle is one of the sturdiest shapes for supporting weight because it distributes stress so well. You can find triangular trusses in some very unexpected places.

To find the reason for the greater strength of the second bridge, carefully peel away the top layer of the second strip of corrugated cardboard. Notice the *corrugations*, little triangular tubes that run across the length of the strip. As you can see, corrugated cardboard (or any other corrugated material) uses a modified truss design for strength. Remember how important it was to cut your strip lengthwise, following these corrugations? What would happen to your bridge if the corrugations ran across the width of the strip?

Display Tip

Exhibit your bridge models along with a clear explanation of the structural principles associated with each model. Collect photographs of various types of bridges—particularly truss bridges—and display them with your models.

Measuring Atmospheric Pressure

You Will Need

- Hanging spring, or grocer's, scale (available in hardware stores)
- Small rubber plunger
- Screw-in hook
- Pail with handle
- Gardening gravel
- Cup
- Pan of water
- Small table
- 4 cinder blocks (optional)
- Poster board
- Pad and pencil
- Ruler

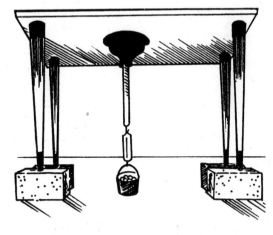

This project will allow you to calculate the *atmospheric pressure* pressing down on the cup of a rubber plunger.

The term *atmospheric pressure* means the weight of the air pushing down on us. We don't feel this pressure because, over the course of evolution, the human body has adjusted to it. Scientists can only measure atmospheric pressure by testing the effect it has on other objects.

Part 1 Plunger Tests

Procedure

1. Hold the cup side of the plunger against the underside of the table. If less than 2 feet (60 cm) remain from the tip of the plunger to the floor, prop up each table leg with cinder blocks or books.
2. Have an adult help you screw the hook into the end of the plunger pole.
3. Dip the plunger cup in a pan of water, and immediately press the plunger against the underside of the table.
4. Attach the grocer's scale to the hook at the end of the plunger pole, and attach the empty pail to the hook on the grocer's scale.
5. Record the weight of the empty pail.
6. Add gardening gravel to the pail 1 cup (240 ml) at a time. Continue watching the weight of the pail until the plunger drops.
7. Record the final weight.

Part 2 Plunger Calculations

Procedure

1. You need to find out the area of the plunger cup before you can calculate the atmospheric pressure acting upon it. On the poster board, trace a circle around the cup of the plunger with the marking pen.
2. Use the ruler and pen to draw a square just large enough to contain the circle. Mark 1-inch (2.5-cm) sections along each side of the square, and draw lines across the square connecting the marks. You should wind up with a grid of 1-inch (2.5-cm) squares over most the circle.
3. Count the number of 1-inch (2.5-cm) squares completely covered by the circle.

4. Count the number of squares partially covered by the circle and give them fractions according to the chart below.

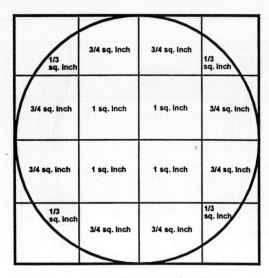

Surface Area

5. Combine the fraction squares so that you come up with a new total of 1-inch (2.5-cm) squares in your grid. This new number represents the real area of the plunger cup.
6. Take the last weight you recorded before the plunger dropped, and divide it by the new total of 1-inch (2.5-cm) squares. Write down the result (quotient) of your division.

Result: The quotient represents *pounds of atmospheric pressure per square inch* on the cup of your plunger. For example, the previous diagram showed a circle divided into sixteen 1-inch (2.5-cm) squares. By combining the fractional squares, you come up with a total area of 11 inches (27.5 cm). If this particular plunger held 42 pounds (19 kg) of gravel before it dropped, divide 11 into 42 for a quotient of 3.8—or practically 4 pounds of pressure per square inch.

In metrics that's 27.5 (the total area in centimeters) divided by 19 (the weight in kilograms) for a quotient of 1.4 kg of pressure per square centimeter.

Explanation

A plunger works because it allows you to squeeze the air from the cup and create a partial vacuum. Although it might seem as if the plunger *sticks* to the bottom of the table, the tremendous pressure of the Earth's atmosphere is actually holding it up. By dipping the plunger in a pan of water, you created a better seal around the edge of the bell so that less air could creep back in. Sooner or later, however, air creeps under the edges of the bell and fills the vacuum. The air pressure inside and outside the bell becomes the same again, and the plunger drops to the ground, pulled by gravity.

Display Tip

Document your experiment with photographs, and exhibit the disassembled parts of your plunger apparatus. Reproduce the grid you used to calculate the plunger's area. Repeat the experiment with various sizes of plungers and record your results. You should know that the results of your experiment will also vary according to your elevation. More atmospheric pressure exists at sea level than at higher elevations. Using sensitive instruments, scientists calculate sea-level pressure at 15 pounds per square inch (6.8 kg per 6.5 square cm).

Did You Know?

The first and probably the most famous of plunger experiments was performed in 1652 by Otto Von Guericke in the town of Magdeburg, Germany. This amateur scientist was also the town's mayor, and so he had very little trouble staging an elaborate demonstration of his vacuum ball in the public square. In fact, the ball was nothing more than two plungers pressed together and emptied of air. But the resulting vacuum was so strong that even teams of horses could not pull the ball apart. The experiment made Von Guericke famous and the German emperor rewarded him handsomely.

Foucault's Pendulum

You Will Need

- 2-liter plastic (soda-pop) bottle with screw cap
- Fine white sand (enough to almost fill bottle)
- Roll of strong nylon cord
- Small, sharp nail
- Sharp pencil
- Hammer
- Scissors
- Masking tape
- Funnel
- Old black or dark-colored sheet
- High folding ladder or access to swing set

In this project, you'll demonstrate the rotation of the Earth by constructing a simplified version of the Foucault pendulum, named after the French physicist Jean Foucault.
We know that the Earth moves beneath our feet. Except for the rising and setting of the Sun, it's difficult to sense this motion. In 1851, Foucault designed an ingenious device that allowed him to demonstrate the Earth's rotation to a fascinated audience.

Procedure

1. Have an adult help you use the sharp nail to punch four evenly spaced holes around the base of the soda-pop bottle. Enlarge each hole with the sharp pencil.
2. Remove the cap from the bottle, and use the hammer and nail to make a hole in the center of the cap. Cover the hole with a piece of masking tape that you can easily remove.
3. Cut two 18-inch (45-cm) lengths of nylon cord from the roll. Carefully thread the first piece through opposite holes in the bottle, and repeat for the second piece. You should have the cords crisscrossing inside the bottle.

Tip: If you have a difficult time getting the cord to thread through both holes, straighten a paper clip and attach the end of a cord to it with a little cellophane tape. Dangle the paper clip through the first hole and maneuver it through the second hole so you can grab it from the other side and pull the cord through the bottle.

Pendulum Detail (steps 1–3)

4. Put one piece of masking tape over each of the holes. Make sure you tape all of the cords in the same position, also pointing toward the bottom of the bottle.
5. Insert the funnel into the mouth of the bottle, and pour in the white sand. Beach sand is fine for this, but make sure you remove any pieces of debris, such as pebbles or plant matter, that might get stuck in the hole. You should have enough sand to almost fill the bottle.
6. Have your adult helper turn the bottle upside down and hold it for you. Tie together the four lengths of cord protruding from the bottom (now the top) of the bottle.
7. Lift your now heavy bottle by the knot and make sure that the bottle hangs *straight down*. Adjust either the knot or length of the cords if it doesn't. Tie nylon cord from the roll to the knot.
8. For your pendulum to clearly indicate the Earth's rotation, you need to suspend it at least 10 feet (3 m). Have your adult helper tie the nylon cord to the center of a folding ladder, if you have one that's tall enough. If not, use a swing set, but make sure you first

pull the swings out of the way. The resting pendulum should hang about 3 inches (7.5 cm) from the ground. Cut the nylon cord and tie it securely when you've hung your pendulum correctly.

9. Spread the black sheet below your pendulum, smoothing it out as much as possible.

10. Grab the pendulum and walk with it, keeping the cord stretched, until the pendulum is about as high as your chin. Remove the piece of masking tape covering the hole in the lid, and release the pendulum.

Pendulum Setup

Result: The pendulum swings, tracing a line on the black sheet. As it continues to swing, the line appears to shift very slightly to the right. But you'll only see this if you can keep the pendulum swinging for at least 15 minutes. *Very carefully* push the pendulum every three complete swings or so, making sure that you push gently and straight so that the pendulum doesn't deviate from its course. Your patience will pay off, since the shift you'll begin to see shows the actual rotation of the Earth!

Explanation

In the Northern Hemisphere, the line shifts to the right because the Earth's clockwise rotation introduces a new force on the pendulum's natural tendency to swing back and forth without changing direction. This natural back-and-forth motion is based on two things: the sideways direction of the first swing, and the downward pull of gravity. Since Foucault's pendulum shifted, he proved that a *third force*—namely the Earth's rotation—was acting upon the pendulum.

The rate of the pendulum's shift depends on how close you are to either the pole or the Equator. At the North Pole, the pendulum would trace a complete circle in about 24 hours. At the Equator it wouldn't shift at all.

Display Tip

Unless your science-fair rules make an exception, your pendulum model will be too large to display in your booth. But you can take detailed photographs showing the construction of your pendulum and how it operates.

Research Foucault pendulums around the world. A pendulum exists at the Smithsonian Institute in Washington, D.C., and in the Convention Center of Portland, Oregon. There's also a wonderful one at the London Science Museum. See if there's a pendulum near your home, and visit it if possible, taking photographs to add to your display.

Did You Know?

Closer to the poles, pendulums more clearly show the Earth's rotation. At the Cathedral of St. Isaac's in St. Petersburg, Russia, a huge pendulum weighing hundreds of pounds (kg) has entertained visitors for years. As the pendulum rotates, it knocks down small pins placed along the sides. Scientists keep the pendulum swinging with the help of powerful electromagnets.

A pendulum swinging in the Southern Hemisphere of the Earth will shift in the opposite direction. This is because the Earth rotates in a counterclockwise direction for people living south of the Equator.

Self-Filling Water Dish for Pets

You Will Need

- Flat-bottom water dish or large disposable broiling pan
- Large jar (without lid)
- Plastic coffee-can cover
- Two 8-inch (20-cm) pieces of 1 × 1-inch (2.5 × 2.5-cm) wood
- Electrical tape
- Ruler
- Water

This is great project for any pet whose water bowl needs regular refilling. Simple atmospheric pressure allows the dish to automatically refill itself from an upright jar. You may need adult help.

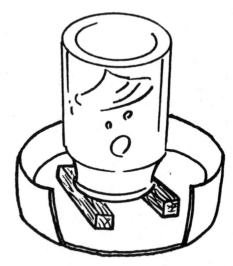

Self-Filling Water Dish *(cutaway view of dish)*

Procedure

1. Before you begin, place the jar and pet dish close together.
2. Use the ruler to measure the diameter of the mouth of the jar.
3. Place the pieces of wood side-by-side in the center of the pet dish, at the same distance apart as the diameter.
4. Attach the pieces of wood to the bottom of the water dish with strips of electrical tape at the edges. Use as little tape as possible.
5. Fill the pet dish until about 4 inches (10 cm) of water covers the wood.
6. Fill the large jar to the rim with water.
7. Place the plastic coffee-can cover over the mouth of the jar.
8. While holding the coffee-can cover against the jar's mouth, have an adult slowly turn the jar upside down. The cover will remain against the bottom of the jar.
9. Carefully move the jar to the pet dish, and set it down so that the cover is pushed under water but doesn't touch the wood.
10. Slowly slide the cover away from the mouth of the jar, and put the jar down on the wood pieces.
11. When the water level in the bowl settles and no more bubbles appear in the jar, carefully strip the tape from the wood pieces.

Result: As you remove the plastic cover and maneuver the jar in place, some water may flow from the jar into the pet dish. But the flow stops when the water level reaches the mouth of the jar, and much more water remains in the jar than in the dish.

Explanation

Atmospheric pressure against the water in the dish keeps the bottle filled. Less pressure exists in the bottle because there is less air, and so the water level inside is higher. But as your pet drinks water from the dish, more air flows into the bottle, and the increased air pressure forces the water level down.

Display Tip

Exhibit the finished water dish. Document the construction of your water dish with photographs, including a photo of your pet using it. Observe and record your pet's reaction to the bubbling water in the jar each time it takes a drink. Does your pet seem to like this new design? Now that you understand the principles involved, do you think you could improve the water dish to make it more pet-friendly?

Cross-Section of Jar and Dish

Escape Velocity

You Will Need

- Piece of PVC pipe, 2 feet (60 cm) long
- Adhesive paper for lining shelves
- Four 1 × 1-foot (30 × 30-cm) square pieces of smooth corkboard
- Pushpins
- White glue
- Marbles
- Modeling clay
- Scissors
- Small handsaw (optional)

One of the most difficult and expensive parts of space travel involves creating a rocket engine powerful enough to escape the Earth's gravity. This project uses easily found materials to simulate what scientists call *escape velocity*.

Procedure

1. Apply glue to the edges of the corkboard squares, and fit them together to form a larger square.
2. Cut a circular piece with a diameter of about 6 inches (15 cm) from the adhesive paper.
3. Carefully pull the backing off the circle, and place the circle (sticky side up) on top of the corkboard and against one edge. Use some pushpins to keep the circle flat against the corkboard.
4. If you have a handsaw, cut one end of the PVC pipe off at an angle. Place that end against the adhesive paper side of the corkboard.
5. Use the clay to build up the end of the PVC pipe so that the inner surface of the pipe is exactly level with the adhesive paper.
6. Prop up the other end of the pipe on a book.
7. Drop a marble into the inclined end of the PVC pipe.
8. Observe the behavior of the marble as it exits the pipe onto the adhesive paper.
9. Raise the inclined end of the pipe and repeat step 7, observing.
10. Raise the inclined end further and repeat step 7, observing.

Result: The first marble exits the pipe but rolls only a short distance on the adhesive paper. The second marble rolls a longer distance but still sticks to the paper. The third marble rolls fast enough so that, although it's slowed by the sticking paper, it continue to roll off the paper and onto the corkboard.

Explanation

At some point, the angle of the PVC pipe will be such that a rolling marble gains enough inertia to pass over the adhesive paper and onto the corkboard. This is because as the pipe's angle of inclination increases, so does the momentum of the rolling marble. Eventually, the marble's *escape velocity* is stronger than the "gravity" of the adhesive paper, and so the marble travels onward into "space."

Display Tip

Document the construction of your model with photographs, and display the finished model. Observe and collect data during your marble rolling experiments, including how each marble behaved as it rolled at various angles. Use a protractor to record the angle at which the first marble escaped the pull of the adhesive paper.

Did You Know?

In the not-too-distant future, scientists from many nations may build a permanent space station. This station won't orbit the Earth like a satellite, but move along with the Earth's rotation. To avoid wasting fuel getting construction materials out into space, some scientists believe that they could build a kind of gigantic electromagnetic elevator. The elevator would move up a cable with one end anchored to Earth and the other end attached to the space station.

Earth & Moon Balance

You Will Need

- Modeling clay
- String
- Scissors
- Pencil
- Ruler

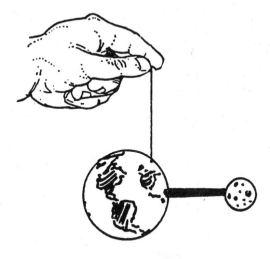

Orbiting bodies abound in space. The Sun moves in a slow orbit about the center of the galaxy, the planets orbit the Sun, and moons orbit the planets. The movement of one body about another raises difficult mathematical problems. Among them is the idea that two orbiting bodies—a planet and moon, for example—have a distinct point (called their *barycenter*) which is the gravitational center of their combined orbits. This project demonstrates this by balancing two clay spheres.

Procedure

1. Cut a 12-inch (30-cm) length of string, and tie it 1 inch (2.5 cm) from the sharp end of the pencil.
2. Mold the modeling clay around the string end of the pencil, adding enough clay to make a baseball-size sphere. Make sure that the sphere sticks out of the pencil and the string comes out of the sphere's edge.
3. Add a marble-size lump of clay to the eraser end of the pencil.
4. Hang your Earth and Moon model by the string.

Result: The pencil should stick straight out horizontally. If the pencil dips in any direction, you should either add clay to the Moon or remove clay from the Earth. When balanced, the Earth and Moon should hang directly opposite each other.

Explanation
The point within the sphere where the string connects to the pencil represents the gravitational center of the Earth and Moon's orbits. This point lies 2,720 miles (4,352 km) beneath the Earth on the side facing the Moon.

Display Tip
Use different colors of clay to represent the Earth's continents and craters of the Moon. Exhibit your model alongside a chart that clearly explains the gravitational center of orbits.

3-D Magnetic Field

You Will Need

- Wide, flat-sided jar with lid
- Mineral oil
- Iron filings
- Tablespoon (15-ml spoon)
- 2 bar magnets

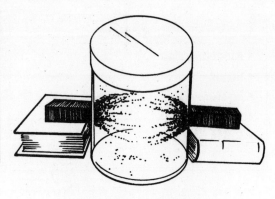

Bar Magnets and Oil-Filled Jar with Iron Filings

It's easy to create a flat representation of a magnetic field using iron filings, a bar magnet, and a white piece of paper. But magnetic lines of force actually move outward in space in many directions and not just along a flat surface. By suspending iron filings in clear mineral oil, this project allows you to see a more realistic magnetic field in three dimensions.

Procedure

1. Fill the jar almost to the top with mineral oil.
2. Add about 3 tablespoons (45 ml) of iron filings. If you still have room at the top of the jar, add more oil. To avoid bubbles, you want as little air in the jar as possible.
3. Screw the lid on the jar securely, and shake the jar.
4. Place the jar on a flat surface, and immediately hold the ends of two bar magnets against the jar, on opposite sides.
5. Turn the magnets around so that the opposite ends touch the jar. Watch the behavior of the iron filings in the mineral oil.

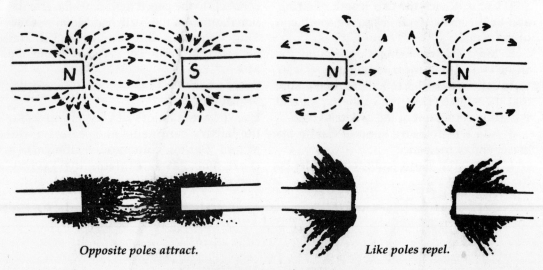

Opposite poles attract.   *Like poles repel.*

Demonstration of Magnetic Fields
Iron filings help illustrate the fields.

3-D Magnetic Field

Result: Shaking the jar caused the iron filings to disperse throughout the oil in a random mixture. But when you held the magnets against the jar, the particles suddenly traced the lines of magnetic attraction or repulsion, depending on how you positioned the magnets' poles. Unlike electricity, magnetism easily passes through many materials, including glass and oil. Suspended in oil, notice how the iron filings move out in all directions to clearly trace the lines of magnetic force. This kind of representation isn't possible on a flat surface.

Display Tip
Allow passers-by to shake the jar and try the experiment for themselves. Use different shapes of magnets; then test and compare the lines of magnetic force between them.

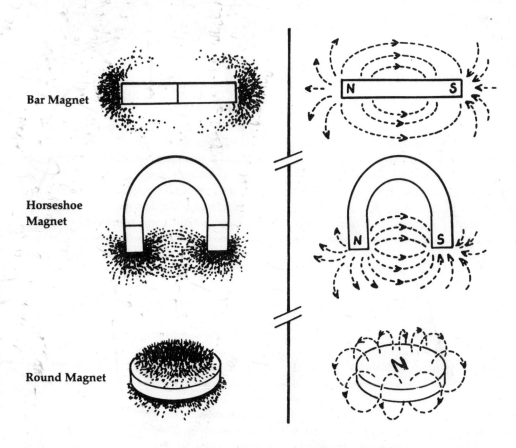

Magnetic Fields of Various Types of Magnets
Iron filings help illustrate the fields.

Homemade Rock Tumbler

You Will Need

- Shoe box lid
- Plaster of Paris
- Bucket
- Plastic trash bag
- Towel
- Hammer
- Glass jar with lid

You can make a homemade rock tumbler with nothing more than a glass bottle, water, and some stones. Rock tumblers help jewelers polish stones by imitating the abrasion of rocks in nature. Abrasion occurs naturally when water rushes against rocks or when particles of rocks are swept along in a current. Abrasion also occurs when wind blows sand against rocks.

It's easier to see abrasion when you use soft rock, like white limestone or sandstone. If you can find samples of these rocks in your neighborhood, break them up and use them. If not, hardened plaster makes a good substitute.

Procedure

1. Mix plaster of Paris and water in a bucket until the mixture gets very thick and hard to stir.
2. Pour the thick plaster in the shoe-box lid, and allow it to harden.
3. Place the lid in the plastic garbage bag, and wrap the bag in a towel.
4. Use the hammer to break up the plaster inside the towel-wrapped bag.
5. Open the bag and remove 25 "stones," all about the same size. Put one stone aside and add the rest to the bottle.
6. Fill the bottle half-full with water, and screw on the lid.
7. Shake the jar 100 times, and remove one of the stones.
8. Shake the jar another 100 times, and remove a second stone.
9. Shake the jar 100 times again, 100 times after that, and 100 times more for a total of 500 shakes, each time removing a stone.
10. Compare all of your stones, including the original stone you saved at the beginning.

Result: The more shakes, the smoother and rounder the stone, so that the last stone you removed doesn't look anything like the original stone. The water in the bottle is cloudy and filled with tiny particles of plaster.

Explanation

Rock tumblers do the work of nature by forcing rocks to grind against each other until they become smooth. As the rocks collide, small pieces of them break off and mix with the water until the water itself becomes part of the abrasion process—a kind of liquid sandpaper. Professional rock polishers leave their tumblers on for weeks or even months at a time to polish hard rocks like granite or quartz. That might seem like a long time until your realize that a month of tumbling represents about 100 years of natural abrasion.

Homemade Rock Tumbler

Display Tip
Arrange your rocks in order of smoothness, gluing each to a piece of construction paper showing the number of shakes. Or, you might want to make a chart showing how different kinds of stones react to 500 shakes.

Experiment with many different kinds of stones, shaking them until they become smooth. But be prepared to shake your bottle for longer periods of time.

Type of Rock \ Number of Shakes	None	100	200	300	400	500
Limestone						
Shale						
Granite						
Quartz						

Sample Chart for Type of Rock and Number of Shakes

Creeping Soil

You Will Need

- Protractor
- Stiff piece of cardboard
- Red string
- Small metal nut
- Pushpin
- Piece of white paper
- Ruler
- Marking pen
- Clipboard
- Rubber cement

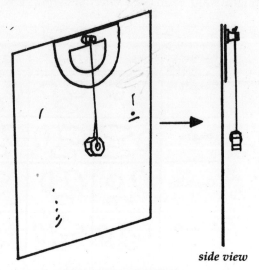

Protractor and Weight

side view

The ground we stand on moves all the time, even when it appears motionless. This is particularly true on a hillside where the soil actually creeps along in a slow downward flow. This project teaches you to measure the amount of creeping soil on a hillside by using a protractor and weight to determine the amount of lean in trees, telephone poles, and other vertical objects. You'll record your results in a chart.

Procedure

1. Glue the protractor to the cardboard so that the flat edge of the protractor is against the top edge of the cardboard.
2. Stick a pushpin in the center hole of the protractor.
3. Tie the metal nut to the end of the red string, and tie the free end of the string to the pushpin. The red string should swing against the protractor, indicating the degree of the angle.
4. Use the ruler and marking pen to divide a sheet of paper into a grid pattern. Glue two strips of paper at the bottom and side edge of your grid so that you can write degrees and the number of objects. Place the grid on a clipboard to take with you when you make your field observations.
5. Find a hill with many vertical objects such as trees, fence posts, and telephone poles.
6. Hold the edge of the cardboard against one of the leaning objects, and note the angle indicated by the red string against the protractor.
7. Record the information on your chart by pencilling in one of the grid squares.
8. Continue to measure leaning objects until you've measured about fifty.
9. Continue recording information until you have a bar graph indication for leaning objects.

Result: Your chart shows, in bar-graph form, how many objects you've measured and their degree of lean. The greater the number of objects leaning, the more the soil is creeping downward under them. The steeper the degree of lean, the more rapidly the soil is moving.

Explanation

With soil creeping, loose materials gradually slide down a slope instead of tumbling suddenly, as in a landslide. The process of slow movement occurs steadily in most regions and is probably more important than landslides in overall erosion.

As you've noticed, examples of creep are everywhere—telephone poles, fence posts, stone walls, and even gravestones will tilt as the soil beneath them moves. Sometimes these objects need resetting every few years to keep them from tumbling over. With trees, sometimes very young saplings adjust to the tilt by bending back up again. But this results in a bent trunk that stays with the tree for the rest of its life.

Display Tip

Have a friend take pictures of you on your field expedition. Exhibit those pictures next to your protractor and completed charts. Make sure you indicate where you took your measurement and any other interesting geologic facts about the area.

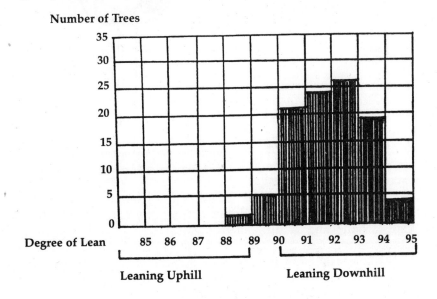

Sample Chart for Measuring Degree of Lean of Selected Trees

Earth Bulge

You Will Need

- Blue poster board
- Rubber cement
- Wooden dowel 1⅝ inches (4 cm) in diameter
- 3 blue pushpins

Although you can't feel it, the Earth spins at a tremendous speed. In fact, a person living at the Equator—where the rotational speed is the greatest—moves about 24,000 miles (38,400 km) each day. Divide this speed by 24 hours and you can see that the rate of the Earth's rotation is about 1,000 miles (1,600 km) an hour!

In northern latitudes, we rotate at about 800 miles (1,280 km) an hour, and at the poles, this rate of rotation is even less.

Although you might think of the Earth as a solid sphere, this force of rotation actually distorts the Earth's shape. Our spinning Earth bulges at its waistline and flattens out near the poles. You can see this by constructing the following model.

Procedure

1. Cut two strips of poster board 22 inches (55 cm) long and 1½ inches (3.75 cm) wide.
2. Make a loop from each strip by gluing the ends together.
3. Fit the loops inside each other, one loop crossing the other, so that together the loops make a kind of globe. Glue the loops at the points where they overlap—the north and south poles of your globe.
4. Apply a little glue to the top of the dowel, then connect the globe to the dowel at one of its poles. Carefully push the three pushpins through the pole and into the dowel.

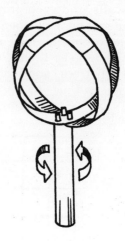

Rotate "globe" by moving your hands back and forth against each other. (step 6)

5. Allow the glue to dry. Then hold the dowel in front of you, and study the shape of your globe.
6. Place the dowel between the palms of your hands, and rapidly move your hands against each other. Watch what happens. Do all parts of the globe seem to rotate at the same speed? Does the globe keep its shape?

Result: The motion of your hands causes the dowel and globe to spin, and the spinning is faster at the globe's "equator" and slower at its "poles." The faster you spin the globe, the more it bulges around the equator and flattens at the poles.

Explanation

The motion of *centrifugal force* changes the shape of the globe as it spins on its axis, the dowel. Centrifugal force is the same force of motion that keeps the water in a pail when you swing it overhead or makes you lean to one side when the car you're riding in turns suddenly. Only the Earth's gravity keeps us from flying off the Earth and into space due to centrifugal force. Can you imagine the power of this force that can cause our gigantic Earth to bulge?

Display Tip
Exhibit your bulging-Earth model along with a detailed description of the Earth's rotation and how centrifugal force affects the Earth's shape. You might want to build several models, some of which viewers may test for themselves.

Did You Know?
Scientists have discovered other forces that affect the Earth's shape. Recently, the U.S. Geological Survey reported something called Earth tides, or a measurable rising and falling of the Earth's *solid* surface in response the gravitational pull of the Sun and Moon. Along the Equator, scientists have measured a 10-inch (25-cm) difference between high and low Earth tides!

Attractive Objects

> **You Will Need**
>
> - Shoe box or small cardboard box
> - Shirt cardboard
> - Marking pen
> - Ruler
> - Bar or horseshoe magnet
> - Collected objects, such as: paper clip, brass paper fastener, piece of chalk, eraser, marble, rubber band, nail, washer, safety pin, key, coin, small battery, ballpoint pen, metal button, soup can, soda-pop can, screw-top bottle cap, toothpick

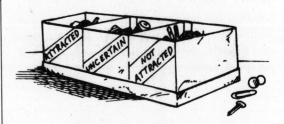

Box with Compartments to Test Objects for Magnetic Attraction

What kind of everyday objects stick to magnets? You might feel pretty sure that cans, nails, and paper clips do. But you're probably not so sure about keys, jewelry, and bottle caps. This project shows you how to collect objects and test them for magnetic attraction. You'll be surprised at some of the results.

Procedure

1. Use the ruler to measure the width of your box. Do the same for the length of the box.
2. Divide the length of the box by three. Mark the top edge of the box to show three equal sections. If your box is 12 inches (30 cm) long, for example, you should make two marks 4 inches (10 cm) apart.
3. Cut the shirt cardboard in half. Take two pieces and trim them so that they're just a little wider than the box. Fold the edges, and test to make sure that each piece fits snugly inside the box. Then glue the edges, and place each piece inside the box at the mark along the top edge. You should wind up with three equal-size compartments in your box.
4. Use the marking pen to label each compartment along the side of the box. Label the first compartment "attracted," the last compartment "not attracted," and the compartment in between "uncertain."
5. Collect the objects on the list and any other objects that interest you. Look for objects, such as jewelry or fancy coat buttons, that combine two or three colors of metal. Place each object in one of the compartments, wherever you think it belongs.
6. Make a list of the objects in each of the compartments, and beside each item on the list give the reasons you think it will attract or will not be attracted, or why you're uncertain about it.
7. Test each item by holding the magnet against various parts of it. Compare your results with your list of objects in the box and your hypothesis for each item. Were you accurate?

Result: Of the objects in your box, it might surprise you that soda-pop cans, keys, coins, and jewelry won't stick to the magnet, but nails, washers, soup cans, and paper clips will. Just by looking at your objects, it's almost impossible to tell which ones will stick to the magnet and which ones will not.

Explanation

Many objects today combine various types of metal. Sometimes a thin layer of a nonattractive metal like brass covers a duller layer of iron or nickel. We call this *plating*. It means that an object like a key may look like it is made of brass but is actually made mostly from a magnetic metal. Scientists have another name for magnetic metals: *ferromagnetic*. The most common ferromagnetic metals are iron, nickel, and cobalt.

How about those partially attracted objects like a ballpoint pen? The clip of the pen (made of steel, an iron alloy) sticks to the magnet, while the body of the pen, made from aluminum, does not. And just in case you have not figured it out already, the "tin" of a soup can is actually a tin coating over steel, which attracts the magnet. The aluminum of the soda-pop can does not react to the magnet at all.

Display Tip

Display your box, objects, lists, ideas, and conclusions. Make a list of larger objects in your home or school that might be plated or might mix nonmagnetic metals with magnetic metals. From a magazine, cut out pictures of common household objects that contain magnetic properties and display them in back of your model.

Ferromagnetic Flowchart

You Will Need

- Large piece of poster board
- Marking pen
- Store-bought bar or horseshoe magnet
- Collection of small objects, such as, key, paper clip, nail, thumbtack, pen, charm, button, wire, and refrigerator pad

This simple flowchart will help you sort out a collection of objects by passing each one through a series of question boxes. Flowcharts provide a useful means of sorting data. In this case, you'll be looking for ferromagnetic properties in your objects, that is, whether or not each object is attracted to a magnet.

Procedure

1. Copy the flowchart onto a large piece of paper or poster board.
2. Take the first object in your collection, and move it into the first circle of the flowchart. If the object appears to be metal, move into the "yes" box and continue to the second circle.
3. If the object doesn't appear to be metal, move it into the "no" box.
4. Hold the magnet against the objects in both the "yes" and "no" boxes. Move all attracted objects into the next "yes" box.
5. Test to see if the magnet attracts all parts of each object. This means turning each object around and holding the magnet against various areas.
6. Separate the objects into the last "yes" and "no" boxes according to what you've found.
7. Examine all the objects in your "yes" and "no" boxes, and determine the common property that makes them either magnetic or nonmagnetic.

Result: No matter how an individual item might look or feel, all the objects in the "yes" boxes contain iron and are *ferromagnetic*. This might also include objects that may not look metallic, such as rubberized refrigerator pads. Objects in the "no" boxes might be made of metal, but not of iron and so not ferromagnetic.

Explanation

It's almost impossible to tell what an object is made of by looking at it. With modern technology, many objects today combine various types of metal. Sometimes a thin

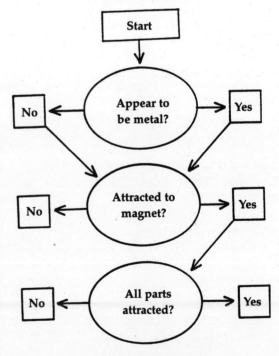

Ferromagnetic Flowchart

What do the objects in these boxes have in common?

layer of a nonattractive metal, like brass, covers a duller layer of iron. We call this *plating*. Partially attracted items are usually made of iron combined with another metal, such as aluminum. Metal buttons often combine aluminum or brass fronts with iron backs.

Display Tip
Display your flowchart, your tested objects, and the results of your flowchart-sifted data. Design a different kind of flowchart to demonstrate the properties of ferromagnetism.

Magnetic Spring

You Will Need

- 5 bar magnets
- 6 sharpened pencils
- Styrofoam "brick"

Can you believe that, someday, cars and trains may float above the ground by using the force of magnetism? You probably think that magnets mostly stick together and attract iron. But the like poles of magnets repel each other with amazing force, and scientists have begun to use that force in practical ways, as this project will show you.

Procedure

1. Lay one of the bar magnets in the center of the Styrofoam brick.
2. With one sharpened pencil, make holes in the Styrofoam around the bar magnet. You should make six holes, two holes against each long side of the magnet and one hole against each narrow side.
3. Remove the magnet, and push a sharpened pencil into each hole. Push the pencils far enough into the Styrofoam to keep them from wobbling.
4. Figure out how to stack the magnets so that the top surface of a magnet always repulses the bottom surface of the magnet above it.
5. Place the first magnet in the pencil support, and drop the next magnet on top of it so that it repels and bounces up. Follow with another magnet until you've stacked all five magnets.

Result: The five magnets will float above each other in the pencil support. Pressing down on the top magnet feels like pressing on a spring, since the magnet will rebound to its original position.

Floating Magnets

Explanation

The magnets float because you've lined up all of their north and south poles on the same side of the stack. The poles, located at the ends of the magnets, have *magnetic lines of force* coming out of them in all directions. So the magnetic force of repulsion caused by like poles acts like a bumper, keeping the magnets apart.

So, if magnets repel each other when you bring their south poles or north poles together and magnets attract when you bring their opposite poles together, why do *both* the north and south pole of a magnet attract iron? It's one of the mysteries of magnetism!

Scientists have long known that the force of magnetism, although related to the force of electricity, is different in many ways. But what really excites them is knowing that magnetism is much stronger than the force of gravity.

In this project, the force of repulsion was enough to defy gravity and keep the magnets floating above each other. This is why inventions like the magnetic spring can be used to **replace** shock absorbers in cars and to help **high**-speed trains glide above ground.

Display Tip
Document each stage of the construction of your magnetic spring. Display the model so that admirers can touch the magnets. From magazines, cut out pictures of common household objects that contain magnets, and display them in back of your model.

The Motion of Gears

You Will Need

- 3 new rolls of paper towels
- 3 new rolls of unwrapped toilet paper
- Cardboard tube from center of toilet paper
- 3 dry-cleaning (trouser) hangers with cardboard tubes
- Colored construction paper
- Cardboard box with shallow lid
- Lids from large and small jars
- Marker
- Scissors
- Cellophane tape

In machines, how can the simple turning motion of different sizes of gears create so many varieties of movement? Gears placed together can spin either toward or away from each other, and each can either slow or speed up the rotation of its neighbor. The following project shows some of the basic principles behind gear motion.

Procedure

1. Cut the box lid so that just the sides remain. This will be the frame for your gear demonstration.

2. Using the marker, draw six dots inside the frame on the long edges, three dots to an edge, facing each other. Draw the first pair of dots in the exact center of the frame and the neighboring pairs 7 inches (17.5 cm) away on each side.

3. Use scissors to punch holes in the dots.

4. Remove three cardboard tubes from the dry-cleaner hangers, and push each tube through the pairs of holes so that you have three "axles" running through the box frame.

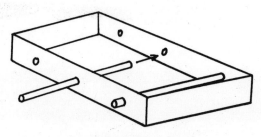

Inserting Axles (step 4)

5. Take the large jar lid and trace three circles on different colors of construction paper. Use the smaller jar lid to trace two smaller circles. Make the smallest circle by tracing the end of the toilet-paper tube.

6. Cut out the six circles and notch them to look like gears. Cut half-dollar size holes in the centers of the three large gears.

7. On each gear, color one tooth black.

8. Tape the three large gears to the edges of the paper towel rolls, tape the two smaller gears to the toilet-paper rolls, and tape the smallest gear to the edge of the toilet-paper tube.

9. Remove the axles from the box frame and place them through the rolls of paper towels. Reattach the axles to the box with the rolls of paper towels (with the gear side facing you). The towels should make three "wheels" in the frame.

10. Place the rolls of toilet paper on top of the paper towels, pyramid-style. Make sure the gears face you.

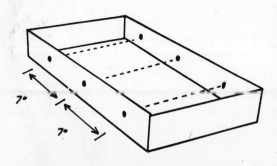

Constructing Gear Frame (steps 1–3)

11. Place the toilet paper tube at the very top.

12. Turn the gears so each gear's blackened tooth is at the 12 o'clock position.

13. Hold the frame and slowly move it across a flat surface. Observe the turning motion of your gears.

Result: If you move the frame to the left, the paper-towel gears rotate to the right. The toilet-paper gears above them rotate to the left. The top tube gear rotates to the right. Although all gears appear to rotate at about the same speed, you can see that the top gear's blackened tooth returns to the 12 o'clock position first, followed by the toilet-paper gears, and then the paper-towel gears.

To add to your experiment, mark out a 6-foot (2-m) course on the floor. Position the gears as before and have three friends nearby to collect data. Slowly move the frame across the floor and have each of your friends watch a different set of gears—paper towel, toilet paper, tube—to record each complete revolution along the course. Compare your totals. By measuring the diameter of the gears, and with a little arithmetic, you can calculate just how much faster each set of gears revolves compared to its neighbor.

Explanation
Gears transfer motion to one another, and engineers carefully figure out what combination of gears in which sizes and shapes will provide the required mechanical action. Sophisticated gear mechanisms turn rotary motion into many other kinds of motion-anything from turning the hands on your clock to lifting a car off the ground.

Display Tip
Display your gear tester along with a description of its operation. You can also display any data you collected while operating your tester, as well as an analysis of that data.

Did You Know?
The first mention of gears occurs in a fourth century B.C. scroll written by a Greek named Stratos, who was a pupil of Aristotle. This scroll is the world's first known engineering text; it describes multiple pulleys and gear-wheel mechanisms, along with other mechanical-advantage devices like the lever and fulcrum.

Gear Demonstration

Puzzling Plant Projects

Phosphates & Algae Growth
Seedy Socks
Colorful Celery Stalk
How Fruits Ripen
Bubble-Gum Plant Graft
Hydroponic Garden
Mushroom Art
Big Green Solar Machine
How Do Leaves Repel Water?
Leaf Rubbings
Cell-Wall Demonstration

Phosphates & Algae Growth

You Will Need

- Three 12-ounce (360-ml) mayonnaise or jam jars
- Measuring spoon
- Masking tape
- Marking pen
- Pond or aquarium water
- Aquarium plants (elodea)
- High-phosphate laundry detergent

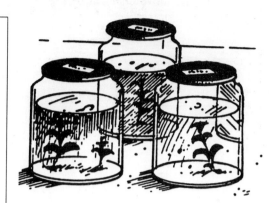

Three Test Jars with Elodea Plants

This project demonstrates the damaging effect of phosphate detergents on freshwater plant and animal life. For the best results, you'll need to collect water containing some algae. Pond water works best for this, but water from an aquarium will do just as well. If you don't own an aquarium, take your jars to a local pet shop and ask someone to fill them with water from their fish tanks. And while you're there, purchase three small elodea plants—your "subjects" for this experiment.

Procedure

1. Place the three water-filled jars near a window that receives a lot of direct sunlight.
2. Label jar #1 "control."
3. Add 1 teaspoon (5 ml) of high-phosphate laundry detergent to jar #2, and label it "one teaspoon."
4. Add 2 teaspoons (10 ml) of detergent to jar #3, and label it "two teaspoons." (If you're using metrics, label the jars in metrics. Jar #2 would be "5 ml," and jar #3 "10 ml.")
5. Place an elodea plant in each of the water-filled jars.
6. Examine the jars carefully over the next two weeks, and record your observations.

Result: In a day or two, the jars with added phosphates will start to turn green, as algae grow and multiply. At first, the elodea plants appear unaffected by the algae surrounding them. In the next few days, however, the phosphate jars become much greener than the control jar, and the 2-teaspoon (10-ml) jar (jar #3) is the greenest of all.

Explanation

Your jars contain over 30,000 different kinds of alga, each kind using the Sun's light to produce the green pigment chlorophyll and to make food through the process of photosynthesis. As the algae continue to grow, they use up all the oxygen in the water, suffocating the elodea. This gradual suffocation by algae is called *eutrophication*, and it occurs because the phosphates in the detergent act as fertilizers for the algae. As the algae multiply out of control, they destroy all other plant and animal life in a pond.

Eutrofication is a serious concern for scientists, since water contaminated with phosphate detergents eventually finds its way into ponds and streams. This results in the death of many useful freshwater plants and organisms.

Use a fork to carefully lift the elodea plant from this jar and examine it. Does it still appear healthy?

Display Tip

Exhibit your living jars with a brief description of their contents. The process of eutrophication will go on for some time, and the algae will thrive until they use up their own supply of oxygen and suffocate.

Seedy Socks

> **You Will Need**
> - Thick cotton socks, almost knee-length if possible
> - 2 rubber bands
> - Shallow bowl of water
> - Magnifying glass

Plants have evolved clever ways of spreading their seeds far and wide. Some flowers, like the fruiting dandelion, have seeds that fly on the wind and scatter for miles. Certain trees, like the maple and elm, have airplane-shape seed pods that flutter to earth far from their sources. But many plants have seeds that simply hitch a ride with anything that passes by—animal, insect, or gym sock.

Procedure

1. Wear high cotton socks and long pants for this experiment. Pull your socks up over the cuffs of your pants and keep the socks from slipping down with rubber bands.

Caution: Never hike through a field with bare flesh exposed. Many insects can harm you, particularly the deer tick that carries Lyme disease, found in many parts of the United States and Canada. These ticks are small (usually poppy-seed size), and naturalists advise wearing light-color clothing to detect them more easily.

2. Find a field of tall grass, and slowly walk through it.
3. When you return home, remove your socks and place them in a shallow bowl of water. Allow the socks to become completely soaked in the water, then move the bowl with socks to a sunny area.

Result: In about a week, your socks will spring to life as many seeds, caught in the fibers of cotton, begin to germinate.

Explanation

Many varieties of grass, such as fescue and cheat, produce seeds with tiny hooks or barbs. This type of seed easily attaches to fibers or fur. Examine some of the seeds in your socks with the aid of a magnifying glass. Although you may find a great variety of seed shapes, all share this hook design.

Display Tip

Your socks, sprouting in their bowl, will certainly attract some attention. Keep a magnifying glass close-by. Use a field guide for your area to help you identify some of the grasses. Make a sketch of some of the more interesting seed shapes, showing some of their common barb structures.

Did You Know?

Seeds are not the only things you or your pet can carry home. *Urushiol*, the toxic oil of the poison ivy and poison oak plants, can stick to clothing and fur and remain dangerous for hours. Always beware of three-leafed plants growing at the base of trees and near water. Wash your pet and your clothing if you're not sure what sort of plants you've stepped through.

Colorful Celery Stalk

> **You Will Need**
> - Long stalk of celery with many leaves
> - Measuring cup
> - Red, blue, and purple food coloring
> - 3 small drinking glasses
> - Scissors

Plants move nutrients throughout their systems through tiny specialized tubes. All plants, from the mighty tree to the humble celery stalk, have this tube system in common. This project helps you see how plants feed themselves.

Procedure

1. Pour ½ cup (120 ml) of water into each of three small glasses. Color the water in the first glass deep blue, deep red in the second glass, and mix the blue and red together for purple water in the third glass.

2. Have an adult use the scissors to split the celery stalk lengthwise into three sections. Place one-third of the split stalk in the blue water, one-third in the red water, and one-third in the purple water.

3. Leave the celery undisturbed for a day or two, and observe what happens.

Result: The celery leaves begin to take on the hues of the blue, red, and purple in three distinct sections.

Explanation

Through millions of years of evolution, the circulatory system of many plants developed into two groups of tubes: *xylem* and *phloem*. Xylem tubes run through the outside layer of stalk or trunk and move mineral nutrients up into the leaves. Phloem tubes are bundled throughout the interior region of the plant and transport food manufactured in the plant's leaves throughout the rest of the plant. In this way, a plant has primitive circulation that keeps it alive.

Display Tip

Since it takes a day or two to achieve the best result, make sure you prepare your celery in time. It's a good idea to prepare several celery samples, in fact, you could start each one a day later than the one before. If a plant begins to wilt during your display, you'll have a fresher one on hand to replace it.

Did You Know?

Damaging the xylem system of a plant can destroy the plant. If you remove a wide ring of bark from a tree, you will starve whatever portion of the tree remains above the ring. For this reason, defacing the bark of a tree is always a serious matter.

How Fruits Ripen

You Will Need

- 2 very ripe bananas
- 3 green bananas
- Unripe avocado
- 3 brown paper bags
- Stapler
- Marking pen

You may have heard the expression "One bad apple spoils the bunch." But you might also say, "One ripe banana ripens the rest," or "One red tomato turns the green ones red, too." In this project, you'll demonstrate how a ripening fruit affects the ripening of other fruits.

Procedure

1. Place the first green banana out in the open air, the second green banana in a brown paper bag, and the third green banana in a brown paper bag with the very ripe banana. Staple the bags closed and label them.
2. Place the first unripe avocado out in the open and the second avocado in a brown paper bag with the other very ripe banana. Staple the bag closed and label it.
3. Leave everything undisturbed for 5 days. Then open the bags and compare all the fruits for ripeness.

Result: The green banana and unripe avocado left exposed show signs of slight ripening in the form of softening and brown spots on their skins. The green banana sealed in the bag shows more ripening, but not as much as the green banana sealed in the bag with the very ripe banana. Here, you'll find that both bananas have turned nearly black. As for the avocado sealed in a bag with the second ripe banana—it, too, shows signs of accelerated ripening when compared to the avocado left exposed.

Explanation

Ripening fruit "breathes" in that it takes in oxygen and gives off carbon dioxide. Oxygen stimulates the ripening process. But, mysteriously, ripening fruit also gives off a gas that speeds the ripening of other fruits exposed to the gas. Known as *ethylene*, scientists call this gas the "ripening hormone."

In your project, the fruit placed with the overripe fruit ripened quickly because of the abundance of ethylene gas in the paper bag. There was also some oxygen in this bag, since small amounts of oxygen can pass through paper. As your banana and avocado combination proved, ethylene gas is a common ripening stimulant for various kinds of fruits.

Although less oxygen was present to stimulate the ripening of the banana left in the bag by itself, the fruit eventually began to breathe its own ethylene and ripen more quickly. Fruits left exposed to the air had plenty of oxygen to help them ripen, but their ripening hormone was lost to air currents and carried away.

Display Tip

Document your experiment by taking photographs at each stage of the procedure. Take good, clear photos of your final results.

Reproduce the setup of your experiment in your booth, displaying the labeled paper bags you used.

Did You Know?
Food manufacturers sometime use ethylene to force greenhouse fruit to ripen, as is the case of the "gassed" tomatoes you buy in the winter. Gassing, however, does not allow a fruit or vegetable's starches to turn into sugars as thoroughly, and so, a gassed tomato will never taste as flavorful as a naturally ripened one. You just can't rush Mother Nature.

Bubble-Gum Plant Graft

> **You Will Need**
>
> - Potted tomato plant about 1 foot (30 cm) tall
> - Potted potato plant, same size
> - Craft knife
> - Soft cotton cord or string
> - Bubble gum

Tomato   "POMATO"   Potato

In this project, you'll graft a tomato and potato plant together. Although you won't truly produce a "pomato," your new plant will still surprise you. Since grafting projects require planning ahead, give yourself at least 8 weeks from the time you graft the two plants to allow your pomato to flower and fruit.

Procedure

1. Pull the main stem of the tomato plant against the main stem of the potato plant. Tie the stems loosely together with string.
2. Carefully shave each stem with the craft knife, just enough so that you expose some of the interior tubes. Ask an adult to help you if you've never used a craft knife before.
3. Press the cut surfaces together, and wrap some string around them, more tightly this time.
4. Pop some bubble gum in your mouth, and chew it until it becomes very soft.
5. Remove the gum from your mouth, and press it around the string you wrapped around the stems.
6. Allow about a week for the graft to set, checking for yellowing and withering on both plants. If the plants look healthy, cut off the top of the potato plant and the bottom of the tomato plant—turning your potato and tomato into a single "pomato."

Result: When tomatoes appear above, carefully dig some of the soil away from the roots of your pomato. You'll see tiny potatoes there!

Explanation

Many of the most familiar fruits and vegetables we eat today did not exist 100 years ago. When botanists (scientists who study plants) began to see how they could combine, or *graft*, plants together, a whole new kind of food was possible. Grafting is a delicate process, and the grafted plant takes many years to produce fruit that combines the

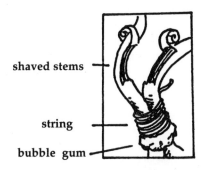

Bubble-Gum Graft of Tomato and Potato Stems (steps 1–5)

qualities of the two original fruits. These new kinds of fruit are called *hybrids*, one of which is the familiar nectarine (peach and plum). Botanists can also graft together two kinds of plum or apple trees to produce sweeter or more sour fruits, or fruits that better resist extremes in temperature.

Display Tip
Exhibit your graft with both tomatoes and potatoes exposed. Document the grafting process with photographs.

Did You Know?
Today scientists are less interested in grafting and more interested in the genetic engineering of plants. By combining similar genes of different plants, many new plants with particular features can be created. For example, genetic engineering has produced strains of berries that resist disease and fungi, and varieties of vegetables that contain a much higher quantity of a particular vitamin. In the future, scientists hope to combine certain types of animal viruses with fruits in order to produce an edible vaccination! This will make it easier and much cheaper to vaccinate people against certain diseases in poorer countries.

How do you tell a fruit from a vegetable, anyway? Here's a good rule of thumb: fruits have seeds on the inside, and vegetables have seeds on the outside. Now how do you feel about that cucumber salad?

Hydroponic Garden

You Will Need

- Potato
- Fresh carrot with leaves
- Grass seed
- Drinking glass
- Saucer
- 2 flat sponges, same size
- Toothpicks
- Aluminum foil
- Liquid fertilizer (available in gardening stores)
- Knife

In the 1980s, the technique of hydroponics was invented by Russian scientists who were trying to find a way to grow plants without soil. Hydroponic cultivation means that the plant's roots sit in a nutrient-rich substitute for soil and can even remain exposed if irrigated with a steady stream of water.

You can demonstrate some simple forms of hydroponic cultivation with this project. You've probably already had some experience with hydroponics without even knowing it. Many tubers and bulbs, like potatoes and onions, root best in pure water where the tender new roots stay protected from corrosive acids and soil pests. This is also true for seeds, such as the avocado and lima bean. Plant cuttings also do well in pure water.

Procedure

1. Cut the potato in half. Then cut the top of the carrot off, near the leaves. Ask for an adult's help if you're not handy with a knife.
2. Choose the half of potato with the most spuds, and press four toothpicks into the potato midway between the cut and the potato's top.
3. Fill one of the glasses with water, and place the potato in the glass so that the cut end sits underwater.
4. Place the glass and potato in a sunny spot.
5. Fill the saucer with just enough water to make a small puddle at the bottom.
6. Put the carrot top in this saucer, and place the saucer and carrot next to the potato.
7. Add 1 drop of liquid fertilizer to the water in the glass and saucer.

8. Wet both flat sponges, and place one on a sheet of aluminum foil. Add 1 drop of fertilizer to the bottom sponge.

9. Sprinkle some grass seed on the bottom sponge, and cover the seed with the second sponge.

10. Wait a few days, and make a note of any changes in your plants.

Result: In about a week, you should see new sprouts growing from both the potato and carrot. If you lift the sponge that contains the grass seed, you'll see that many of the seeds have sprouted and begun to root in the bottom sponge.

Explanation
The tender new roots are well-nourished by the liquid fertilizer and plentiful water. Eventually, however, your potato, carrot, and grass will need more nutrients and nitrogen than the water and liquid fertilizer can give them. Their roots will begin to crowd and tangle together. At this point, they should be replanted in soil.

Display Tip
Show off your hydroponic garden at several stages, including the stage where the plants show obvious need for soil potting. Can you see the challenge scientists have in keeping mature plants healthy without soil?

Did You Know?
Not quite the same as hydroponic gardening, the term *xeriscaping* means landscaping with low-water plants. Xeriscaping uses drought-resistant plants and low-maintenance grasses that require water only every two or three weeks. This type of gardening has become very popular in dry climates where the use of large quantities of water for gardening is forbidden.

Mushroom Art

> **You Will Need**
>
> - Store-bought fresh mushrooms
>   (DON'T PICK MUSHROOMS FROM THE WILD. SOME ARE POISONOUS!)
> - White sheet of paper

Have you ever seen a mushroom seed? As hard as you might look for one, mushrooms and other *fungi* have no use for seeds. Instead, plants like mushrooms and ferns reproduce by means of *spores,* which act as a kind of primitive plant embryo.

One mature mushroom will produce several million spores from its underside, or gills. These spores are much too tiny to see without the help of a microscope. But you can easily recognize the presence of spores by the patterns they create, as this project will show you.

Procedure

1. Carefully break off a few mushroom caps from their stems.
2. Place the caps, gill-side down, on the white sheet of paper.
3. Leave the mushrooms undisturbed for a few days.
4. Remove the mushrooms from the paper.

Result: The mushrooms left graceful fan-shaped patterns on the paper, traced in brown.

Explanation

As the mushroom caps sat undisturbed, they "ripened," and thousands of sticky spores dropped from the gills onto the paper. These spores accumulated to form a detailed picture of each mushroom's underside. The longer you leave a mushroom undisturbed, the darker the pattern, but even a mushroom left for just an hour will make a faint picture.

Display Tip

Show off your mushroom art, framed if necessary. Try making several mushroom "exposures" on the same sheet—leaving several mushroom on the paper for days and others for only a few hours. You'll see an amazing variety of color and detail in your mushroom art.

Did You Know?

Spores show an amazing variety of design. Besides the sticky variety, some can actually swim when released from the plant over water. Others launch themselves from their spore cases when the outside temperature is just right.

Big Green Solar Machine

You Will Need

- Tree with large leaves (make sure you can reach some branches)
- Piece of white paper
- Ruler
- Pencil
- Calculator

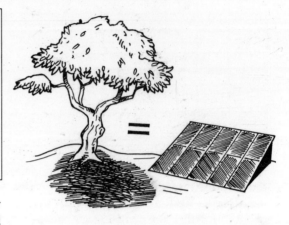

Plants feed themselves by using the energy of the Sun to make food. This process, called *photosynthesis*, means that sunlight helps the plant turn water in the soil and carbon dioxide in the air into sugars.

You can think of the leaves of a plant or tree as miniature sunlight-collectors, or solar panels. If you could figure out the area of all the leaves on a single tree, how large a solar panel do you think you would have? This project helps you find out.

Procedure

1. Use the ruler and pencil to draw a ½-inch (1.25-cm) grid on the sheet of white paper. You might want to make several photocopies of your grid so that you can do this project more than once.
2. Remove a leaf from the tree (or take a fallen leaf from the ground), and place it over the grid.
3. Trace the leaf's outline on the grid, and remove the leaf.
4. Look at how the leaf shape fills the grid. Draw a checkmark in every square half- covered or more by the leaf. Draw a minus sign in the squares that have just a small portion of leaf. Ignore the remaining squares.
5. Add up the checkmarks, and write that number in the left corner of the grid.
6. Observe your tree again. You might need a pair of binoculars to help you. Estimate the number of leaves on a small twig, and write that number down.
7. Estimate the number of twigs on a branch, and multiply this number with the previous number.
8. Estimate the number of branches on a limb, and multiply this number with the former number.
9. Estimate the number of limbs on the tree trunk, and multiply again. This will give you a reasonable estimate of the total number of leaves on the tree.
10. Multiply the total number of leaves with the number you wrote in the left corner of the grid.
11. Divide this number by 288.

Result: The answer, or *quotient*, gives you the total number of square feet the leaves of your tree represent. Here's an example: say you count five checkmarks for your leaf. Write the number 5 in the left corner of your grid. Then you go outside and look at your tree again, and you estimate that there are twelve leaves on a twig and six twigs on a branch. Multiply 6 with 12 to get 72, or 72 leaves on a branch. Next, you estimate five branches on a limb. Multiply 5 with 72 to get 360, or 360 leaves on a limb. And a final estimation gives you four

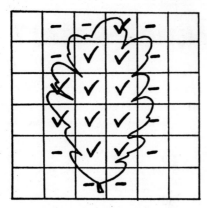

Leaf Grid with Checkmarks
For Estimating Leaf's Surface Area (steps 3–5)

limbs to the trunk. Four times 360 equals 1,440 leaves for the entire tree.

Now for the really big number. Multiply 1,440 by the number 5—the number of checkmarks representing a leaf's area. You should divide your total, 7,200, by 288 (the number of square ½ inches in a square foot) for the quotient 25. This means that the leaves of your tree translate into a solar panel of 25 square feet.

If you want to figure the number in metrics, consult the Metric Equivalents chart (p. 201). Ask your math teacher for help.

Explanation

Leaves and a solar panel resemble each other in that both turn the energy from the Sun into a different kind of energy. With solar panels, the material of the panel, often silicon, turns the Sun's energy into electrical energy. With leaves, photosynthesis turns sunlight into a chemical energy which, in turn, changes starches into sugars. But a leaf is a much more efficient energy factory than a solar panel. Solar panels can only turn a small fraction of the sunlight into electrical energy, while a leaf uses nearly every bit of it!

Display Tip

Calculate the square footage of several trees and measure out, either with chalk on the ground or by using a large blanket, how large a solar panel you'd have for each tree.

How Do Leaves Repel Water?

You Will Need

- Piece of green acetate (available in art supply stores)
- 4 bendable plastic straws
- Piece of shirt cardboard
- Cellophane tape
- Scissors
- Eyedropper
- Bowl
- Water

Plants need water, of course. But plants also need to protect themselves from the *weight* of water as it falls on their leaves. This project shows you how a leaf's design is perfectly suited to do that.

Procedure

1. Cut the four shapes shown on page 177 from the green acetate
2. Bend the four straws, and tape the short ends to each shape. Make sure the opening of each straw is in the middle of the shape.
3. Bend and tape the shirt cardboard to make a stand (see p. 177).
4. With tape, attach the long end of each straw #1 to the cardboard so that the top of the straw pokes out over the cardboard's edge and the acetate shape sticks out from the cardboard. Repeat this procedure for straws #2–4.
5. Place an empty bowl under the shapes.
6. Fill the eyedropper with water, and squirt water down through straw #1. Notice the pattern of water as it flows over the shape, and notice how the shape begins to bend.
7. Squirt water down the remaining straws, each time noticing how the water flows over the shape.

Result: With straws #1 and #2, the water spreads out over the shape in all directions. Soon the shape begins to bend from the weight of the water on its surface. If these shapes were leaves, what do you think would happen?

The shapes attached to straws #3 and #4 do much better. Here, the water flows out toward the tips of the shape so that it can run off and keep the shape from bending. Can you see why these shapes appear more leaflike?

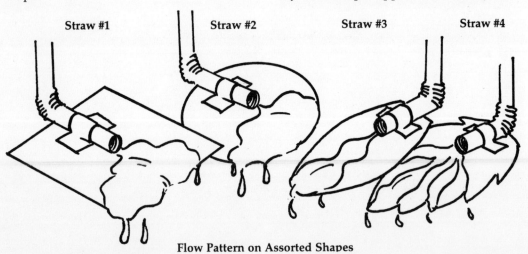

Flow Pattern on Assorted Shapes

How Do Leaves Repel Water?

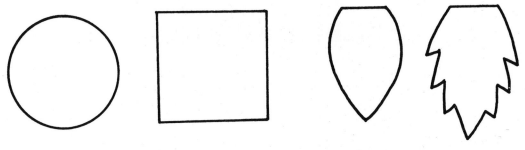

Cut these shapes from green acetate. (step 1)

Explanation
A leaf's design helps it channel water away from its surface. The more points on a leaf, the more "drain pipes" it has to slide the water off the surface. A square of circular leaf would snap with the weight of water, and the leaf would die.

Display Tip
Collect real leaves and exhibit them next to your model. Can you find any square or circular leaves? You may, in fact, find circular leaves on some tree species—but examine them closely. Even leaves that may appear circular actually have small points for the water to run off.

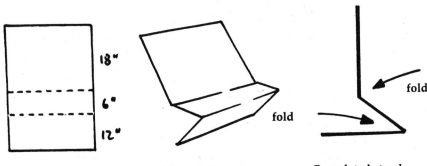

First, measure paper. Dotted lines indicate fold lines.

Then, fold paper along dotted lines.

Completed stand (side view)

Making the Cardboard Stand (step 3)

Leaf Rubbings

You Will Need

- Drawing pad
- Crayon
- Pocket scissors
- Garden gloves

You can more easily see the surface details on a leaf by placing it under a thin sheet of paper and lightly rubbing a crayon against it. Look for interesting leaves around your neighborhood, and choose leaves with large veins or ribs. Before you go out to collect leaves, learn to recognize and avoid leaves you *should not touch*—poison oak, poison ivy, and poison sumac. The leaves of these plants are arranged in groups of three around a single stem. Find a field guide for your area to help you in your leaf collecting.

Procedure

1. Take the scissors and drawing pad with you on your leaf-collecting trip. Cut interesting leaves from trees, bushes, and garden flowers (with permission, of course), and place the leaves inside the drawing pad to keep them flat.

2. At home, prepare a smooth surface where you can do your leaf rubbings. Tear a sheet of paper from the drawing pad, and place it in front of you.

3. Place the leaf, upside down, under the paper. Placing it upside down will give you better results because the veins are more prominent on the underside of a leaf.

4. With two fingers of one hand, press the paper tightly around the leaf. With the other hand, rub the crayon over the paper in even strokes. Don't press too hard.

5. Repeat the procedure with all the leaves in your collection, using different colors of crayons.

Result: Your leaves will leave beautiful images of themselves on the paper. You should see much surface detail, such as edges and ribs—even some blemishes on the leaves from insects and fungi.

Explanation

The crayon highlights the textures of your leaf samples—details you wouldn't ordinarily see. A leaf's surface has many complex structures and textures. In some leaves, the ribs turn into a fine network of veins that fan outward to the very tip of the leaf. This

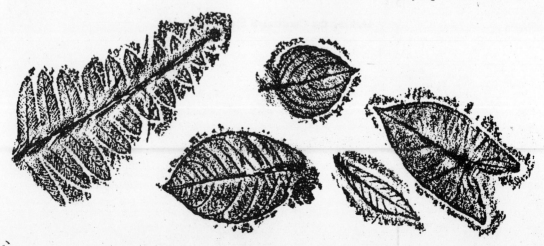

is called *branched venation*. In other leaves, the ribs look like stripes and do not branch. This is called *parallel venation*. Leaves also have tiny structures called stomata which allow the leaf to breathe.

Display Tip
Besides being a scientific record of your leaf specimens, your rubbed leaf designs make beautiful artwork. Display your rubbings in frames and identify leaves with branched and parallel venation and place them in two groups. Do plants of each group have something in common?

You could also identify local trees by their leaves, consulting a tree guide for your area.

Did You Know?
The State of Michigan has more species of tree than any other state in the U.S.

Cell-Wall Demonstration

You Will Need

- Small plastic bag with twist fastener
- Cornstarch
- Clear glass bowl
- Measuring cup
- Teaspoon (5-ml spoon)
- Iodine

This project shows the amazing property of *semipermeability*, or the ability of a barrier to allow only a certain size of molecule to pass through it. Cells, in both plants and animals, have semipermeable walls that allow cells to both feed themselves and remove waste. At the same time, these walls protect cells from substances that might otherwise harm them.

Procedure

1. Add 1 teaspoon (5 ml) of cornstarch to 1 cup (240 ml) of cold water. Stir the water until the cornstarch dissolves.
2. Pour the mixture into the small plastic bag, and twist the fastener around the bag's open end.
3. Pour 4 cups (960 ml or 1 liter) of warm water into the glass bowl. Add just enough iodine solution to turn the water a light yellow.

   Caution: IODINE CAN STAIN HANDS AND CLOTHING. ALTHOUGH IT OCCURS NATURALLY IN FOOD, PURE IODINE CAN BE HARMFUL IF SWALLOWED.

4. Place the bag in the bowl, and swish it around for a minute. Observe what happens.

Result: The water inside the bag first turns blue, then black. The yellow iodine water does not change color.

Explanation

Iodine and starch react chemically to each other. Iodine, mixing with a starch solution, will turn the solution a bluish-black color. Since the starchy water in the bag turned this color, it means that the iodine penetrated the plastic and entered the bag. However, the iodine water in the bowl remained unaffected. This means that iodine molecules *passed through* the plastic bag to react with the starchy water, while starch molecules remained in the bag.

Since starch is one of the largest molecules, it was too big to pass through the semipermeable "cell wall" of the plastic bag. The simpler iodine molecules had no such problem.

Display Tip

Document the dramatic results of this experiment with photographs. Place the bowl and its dark-color bag on display for others to examine. Do not let them touch the display, however.

Fly, Float & Sink

Kite-Sighter
Baffle Test for a Glider
Depth Indicator
Hydrodynamic Hull Designs
Why Is a Kite Like a Sailboat?
Whirligigs & Parachutes
Boat-of-Clay

Kite-Sighter

You Will Need

- Square piece of foam board, 12½ inches (use 40 × 40-cm for metric version)
- 26 straight pins (sewing pins)
- 1 pushpin
- Yardstick or meterstick
- Strong black thread
- 2 small fishing weights
- Black marking pen (sharp point)
- Green and yellow modeling paint
- Red and blue marking pens
- Scissors

This simple but accurate tool lets you determine the height of any stationary object, including a kite, to a distance of 250 feet (about 75 m). The scientific name for a kite-sighter is *clinograph*.

Part 1 Constructing a Clinograph

Procedure

1. Use the yardstick and marking pen to draw a grid on the foam board. Divide the edges of the foam-board square into twenty-five ½-inch segments with vertical lines. Then draw twenty-five horizontal lines so that you wind up with a grid of 625 ½-inch squares. Each ½-inch square represents 10-foot increments of altitude.

If you prefer metrics, for this step you'd use a meterstick to create the grid. Divide a 40 × 40 cm square foam board into a grid of 400 2-cm squares by drawing twenty vertical lines and twenty horizontal lines, with each 2-cm square representing 3-m increments of altitude.

2. Stick 26 straight pins (21 pins for metrics grid) as close to one edge of the square as you can. Except for the pins at the top and bottom of the square, you should place each pin at the end of a grid line.

3. Cut the strong black thread into two 3-foot (1-m) lengths. Paint one fishing weight yellow and the other green. After the fishing weights dry, tie one weight to one end of each thread.

4. Hold the foam-board square so that the row of sewing pins is to your right. The top

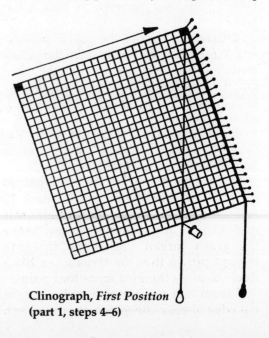

Clinograph, *First Position* (part 1, steps 4–6)

edge of the square is your sighting edge, and you should make this edge stand out from the rest of the grid. Use the red marking pen to color in the leftmost square of the top edge, and use the blue marking pen to color the rightmost square of the top edge.

5. Tie the free ends of thread to the sewing pin at the top of the blue square.

6. Stick the plastic pushpin into one edge of the board, someplace where you can easily reach and remove it.

Part 2 Using a Clinograph

Procedure

1. Take your clinograph outside to test it on a building or any structure the height of which you already know.

2. Hold the clinograph in a vertical position so that the row of pins points away from you. Let the thread with the green weight hang outside the row of pins, and let the thread with the yellow weight hang freely against the grid.

3. Look along the top edge of the clinograph (from the red to blue square), and point to the object you want to measure. Hold the clinograph very steady. The yellow thread will hang somewhere against the grid.

4. Carefully remove the plastic pushpin and reinsert it at the point where the yellow thread passes the bottom edge of the clinograph. Try to hold the thread in position with the pushpin. This marks your first measurement.

5. Walk forward about 150 feet (46 m). You have to carefully measure your paces in order to get an accurate calculation.

6. From the top of the clinograph, count down one sewing pin for each 10 feet (about 3 m) you walked—in this case 15 pins for 150 feet (46 m). Move the green thread under the fifteenth sewing pin so that the thread can hang freely against the grid from that point.

7. From your new position, look along the top edge of your clinograph at your object.

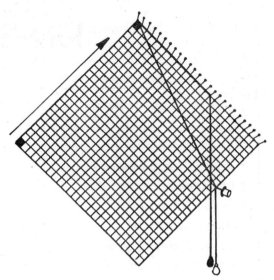

Clinograph, *Second Position*
(part 2, steps 1–8)

Hold the green thread in place at the bottom edge of the sighter.

8. Turn your clinograph so that you can see where the yellow and green threads cross.

Result: The intersection of the threads against the grid enables you to figure out the height of the object. Do this by counting left along the grid from the row of sewing pins until you meet the crossing strings. For example, if the strings cross in the middle of the eighth square, the altitude of the object you measured is 85 feet (25.5 m). Compare your measurement with the known height of the object.

Explanation

The sighter uses a principle of geometry called *triangulation* to calculate the distance to the object. Determining an object's distance by measuring its apparent movement, or displacement, between two different observer points is also called the *parallax effect*. Astronomers use the parallax effect to calculate the distances from Earth to planets and stars.

Display Tip

Document the construction and operation of your clinograph with photographs. Display the finished model next to a chart of the various objects you've measured. List any problems you might've had when first using the clinograph. These might include difficulties in holding the clinograph steady or inaccurately pacing off your distance between sightings. Suggest ways you might improve the basic clinograph design.

Baffle Test for a Glider

You Will Need

- Large piece of poster board
- 3 × 5-inch (7.5 × 12.5-cm) piece of wood, 20 inches (50 cm) long
- 3 × 5-inch (7.5 × 12.5-cm) piece of wood, 12 inches (30 cm) long
- Small sheet of balsa wood
- Four 6-inch (15-cm) strips of balsa wood
- 3-inch (7.5-cm) strip of balsa wood
- Shirt cardboard
- Wire hanger
- Thumbtack
- Rubber band
- 2 straight pins
- Masking tape
- Wire cutters
- Pliers
- Hand drill
- Ruler
- Scissors
- Craft knife
- Rubber cement
- White glue
- Small low-power electric fan

Part 1 Baffle Building

Procedure

1. Cut strips for the baffle's frame and slats from the poster board, following the dimensions in the illustration below.

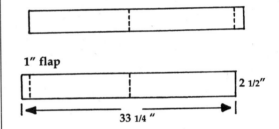

Strips for Baffle Frame (part 1, step 1)

2. Divide both halves of the frame into sixteen 1-inch (2.5-cm) sections, using the ruler and pencil to mark the sections with lines.

3. Fold the two sides of the frame together, and connect them by applying rubber cement to the 1-inch (2.5-cm) flaps.

4. Divide each of the fourteen slats into eight 2-inch (5-cm) sections between the flaps. Draw lines separating these sections, extending the lines no more than halfway through the width of the slat.

5. Use scissors to cut along the lines so that each slat becomes grooved.

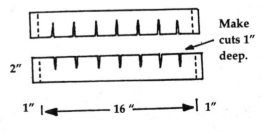

Slats for Baffle (part 1, steps 4–5)

6. Place the frame on a flat surface, and fold back the flaps on each slat.

This project demonstrates how the movable flaps on an airplane's wings and tail section control basic lifting and turning motions. We'll demonstrate these movements by performing a simple wind test on a balsa-wood glider. For more accurate results, a poster board baffle helps "straighten out" the wind that is produced from a low-power electric fan.

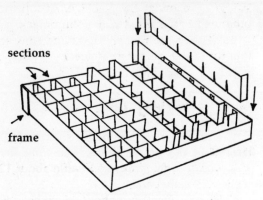

Push slats down into frame.

sections

frame

Baffle Building (part 1, steps 7–11)

7. Carefully insert eight slats into the frame, following the guidelines. You'll notice that the frame is slightly deeper than the slats. Make sure that each slat has its grooves pointing up.

8. Attach each slat to the frame by applying rubber cement inside the flaps.

9. Now insert eight more slats with the grooves pointing down. Place each new slat on top of the attached slat perpendicularly so that the grooves interlock.

10. Attach this new set of slats to the frame by applying rubber cement inside the flaps.

11. For corner braces, use scissors to trim the ends of each 6-inch (15-cm) balsa-wood strip. Use the white glue to secure the strips in the corners of the frame. Allow at least an hour for the glue to dry.

12. Turn the finished baffle on its edge, and attach it to the 20-inch (50-cm) piece of 3 × 5-inch (7.5 × 12.5-cm) wood with two strips of masking tape.

Part 2 Constructing the Stand & Glider

Procedure

1. Use the wire cutters to cut a straight piece of wire from the metal hanger.

2. Use pliers to bend the wire into an L-shape, and make a little hook at the end of the short part of the L.

3. To cross-brace the wire, notch both ends of the 3-inch (7.5-cm) balsa-wood strip, and insert it into the crook of the L. Attach the ends with some masking tape.

4. Take the 12-inch (30-cm) piece of 3 × 5-inch (7.5 × 12.5-cm) wood, and drill a narrow hole near one end. Insert the long end of the wire into the hole so that the L is upside down.

5. Cut one of the rubber bands so that you have a long elastic piece. Tie one end of the rubber band to the midpoint of the top edge of the stand, and attach the other end to the wood with a thumbtack.

6. Use the craft knife to cut the shape of a glider fuselage from the thin sheet of balsa wood. Have an adult help you if you've never used a craft knife before.

7. Cut a small slot in the tail section of the fuselage.

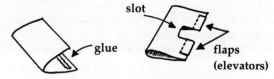

Tail Section (part 2, steps 7–8)

8. Cut a wing and tail section from the shirt cardboard, following the diagram. Cut large flaps (ailerons) in the wing and smaller flaps (elevators) in the tail section, following the diagrams above and below.

9. Use the craft knife to make a lateral slit in the fuselage's midsection, wide enough for the wing to slip through

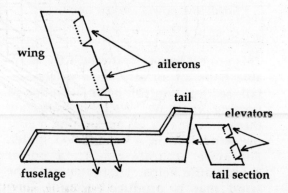

Putting the Glider Together (part 2, steps 9–13)

10. Anchor the wing to the fuselage with two straight pins. Insert one pin into the top edge of the fuselage and the other pin into the bottom edge.

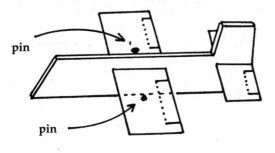

Placing Pins in Glider (part 2, step 14)

11. Attach the tail section by sliding it into the slot you cut in the balsa wood.
12. Cut the rubber band so that you have a long elastic piece. Cut this piece into two shorter pieces so that one piece is about half as long as the other piece.
13. Tie one end of the shorter piece to the hook you made in the wire frame. Tie the other end to the middle straight pin over the wing.
14. Tie one end of the longer piece to the straight pin under the fuselage. Attach the other end to the wood with a thumbtack.
15. Adjust the rubber bands and pins so that the glider sits perfectly level.

Part 3 Wind Test

Procedure

1. Position the glider and baffle about 12 inches (30 cm) apart.
2. Place a small fan about 16 inches (40 cm) behind the baffle.
3. Straighten the ailerons on the wings, and tilt up the elevators on the tail section.
4. Turn on the low-power fan and watch the glider.
5. Turn off the fan and tilt down the elevators. Resume the wind test.
6. Straighten the elevators, and tilt the ailerons on the wing in opposite directions. Turn on the fan and watch.
7. Combine any and all of the above aileron and elevator positions, observing how the glider behaves each time you turn on the fan.

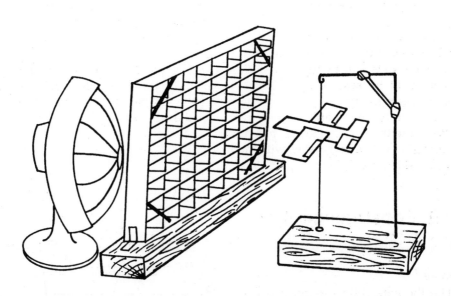

Fan, Baffle, and Glider on Stand (part 3, step 7)
In Position for Wind Test

Result: With the elevators tilted down, your glider should dip as if trying to land. This occurs because the dipping elevators change the motion of air across the wing and tail so that their undersides no longer have the same amount of lift. With the elevators tilted up, the nose of your glider should ascend. This time, the changed motion of air gives the glider more lift. With the elevators straightened and the wing's ailerons tilted in opposite directions, the glider should dip left or right as if preparing to turn.

Explanation

Engineers design many different kinds of wing and tail flaps to help pilots fine tune the climb, descent, and turns of their aircraft. The tail section of an airplane controls climbing and descending. The tail's elevators move up or down when the pilot pulls back or pushes forward a control column or stick near his seat. When he pushes the stick forward, the elevators swing downward and the nose of the airplane dips downward. When he pulls the stick back, the elevators swing upward and the nose of the airplane points upward.

The ailerons on the airplane's wing have less to do with lift and descent but control the banking (turning) motion. With the left aileron moved up and the right aileron moved down, the airplane banks to the left. With the left aileron moved down and the right aileron moved up, the airplane banks to the right. Although your glider doesn't include one, airplanes also have a rudder in their tails to help make banking and turning more stable.

Display Tip

Exhibit your working model for judges and observers to admire. Demonstrate how dramatically the position of elevators and ailerons on the wings and tail change the way a glider behaves. For an added touch, construct a simple movable "stick" by placing a pencil in a lump of clay. Show the different positions of the stick when the glider climbs or descends.

Depth Indicator

You Will Need

- Large pail or empty aquarium
- Small funnel
- Long piece of rubber tubing
- Clear plastic drinking straw
- White poster board
- Balloon
- Rubber band
- Masking tape
- Rubber cement
- Scissors
- Eyedropper
- Red-colored water

For this project, you'll construct a device that measures water pressure at different depths and displays results by means of a tube of colored water. In principle, your model will be very similar to the sophisticated depth gauges used on ships.

Procedure

1. Cut the balloon in half, and stretch the rubber over the wide end of the funnel. Use the rubber band to hold the balloon in place. This is the sensitive diaphragm part of the indicator.

2. Attach one end of the rubber tube to the spout of the funnel. Attach the other end of the tube to the plastic straw.

3. Fold the poster board so that it makes an upright stand, following the diagram. Glue the poster board together at the base of the stand.

4. With the stand sitting upright, tape the tube and straw in a U-shape against the poster-board stand; then temporarily remove the funnel from the tube.

5. Using the eyedropper, half-fill the drinking straw with red water. Replace the funnel on the tube.

6. Add water to the pail or aquarium until about three-quarters full. Push the funnel, rubber-side down, under the water, and observe the red water in the plastic drinking straw.

Result: As you push the funnel deeper into the water, the red water in the drinking straw rises in direct proportion.

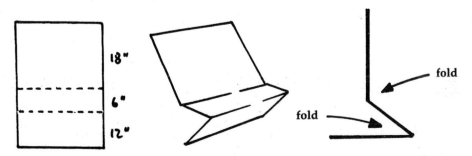

Constructing the Stand (step 3)

Explanation

As the funnel descends, the weight of the water above it increases. This means that increased water pressure pushes against the diaphragm of the funnel. As the diaphragm pushes in, it pushes the air in the tubing, which in turn pushes the red-colored water up into the drinking straw.

Display Tip

Exhibit your working model and demonstrate it for the judges. Try to use an aquarium for the water container since you can more easily see the relationship between the sinking funnel and rising water in the straw.

Hydrodynamic Hull Designs

You Will Need

- Cardboard box with shallow lid
- White plastic garbage bag
- Rubber cement
- Rubber cement thinner
- Cellophane wrap
- Masking tape
- Scissors
- 5 sheets of balsa wood, 1-inch (2.5-cm) thick
- 2 small corks
- Craft knife
- Ruler
- Waterproof marker
- Graph paper
- 2 metal hangers
- Pliers
- Wire cutters
- Sheet of fine sandpaper
- Plastic basin (that you can discard afterwards)
- Drill
- Large water pail
- Water, of course

Closely related to aerodynamics, or the study of air friction against moving objects, *hydrodynamics* helps engineers create sleek designs for ships. A streamlined ship hull means speed and energy efficiency. Here you can test boat hull designs to find the most efficient.

The water sluice in this project serves the same purpose as the wind baffle in the glider project.

Part 1 Making a Sluice & Grid

Procedure

1. Remove the lid of the paper box and turn it upside down. Remove one narrow side, and cut away a section of the other narrow side, following the diagram below.

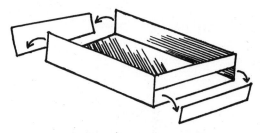

Preparing Box Lid (part 1, step 1)

2. Cut along the seams of the white plastic garbage bag until you have a flat sheet of plastic. Measure the plastic and trim it so that it covers the inside of the lid and both long edges.

3. Remove the plastic, and use a thinned solution of rubber cement to coat the lid.

Replace the plastic and smooth it against the lid. Use the ruler to tuck the plastic tightly into the corners.

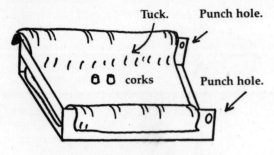

Preparing Sluice (part 1, step 3–5)

4. Put some rubber cement on the wide ends of the corks, and glue the corks to the plastic bag surface. Place the corks about 4 inches (10 cm) apart and slightly below the center of the lid.

5. Punch two holes in the lid, following the diagram.

Cellophane Stretched across Lid
(part 1, step 6)

6. To make the grid, turn the lid hollow-side down, and stretch a sheet of cellophane across it. Leave about 5 inches (12.5 cm) at the top of the lid uncovered by cellophane.

Grid in Place (part 1, steps 7–10)

7. To keep the cellophane stretched, tape the edges to the side of the lid.

8. Mark the long and short edges of the cellophane into 1-inch (2.5-cm) segments, using the ruler and marker.

9. Using the ruler and waterproof marker, carefully draw lines connecting the marks from long edge to long edge, and from short edge to short edge. You should wind up with a 1-inch (2.5-cm) square grid, drawn directly on the cellophane.

10. Carefully remove the cellophane and turn the lid over. Stretch the cellophane over the open side of the lid, taping it as before. Make sure that the end of the lid where you punched holes remains uncovered by cellophane and that you can easily reach the corks under the cellophane.

Part 2 Constructing the Hulls

Procedure

1. Have an adult cut the hull designs shown on the bottom of page 193 out of the sheets of balsa wood with the craft knife.

2. You might want to have some extra sheets of balsa wood around to experiment with some hull designs of your own.

3. Sand the edges of your balsa-wood models smooth.

4. Use the ruler to measure the distance (about 4 inches or 10 cm) between the two corks you glued to the sluice, and cut holes in the middle of each model just large enough for the corks to fit through.

Part 3 Preparing the Basins

Procedure

1. Setting up the sluice requires a small table or low stool for the top basin. First, prepare the top basin by drilling two rows of $1/16$-inch (0.15-cm) holes, regularly spaced, in one of the sides.

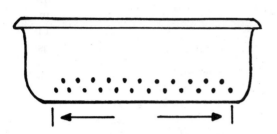

Basin with Holes Drilled in One Side (part 3, step 1)

2. Use the ruler to measure from the lip to the bottom of the basin, and cut two pieces of wire from the wire hangers slightly longer. Use the pliers to bend both ends of each wire into hooks; then hook both wires over the lip of the basin on the side where you've drilled holes.

3. Place the basin on the edge of the stool with dangling hooks facing you. Connect the sluice to the top basin by inserting the hooks through the holes. You may have to make some adjustment in the lengths of the hooks to make sure that the sluice fits snugly under the basin.

4. Rest the unattached side of the sluice on the edge of the second basin, placed on the floor.

Part 4 Testing Hull Designs

Procedure

1. Before you begin your experiment and collect data, you must first reproduce the grid of your sluice on graph paper. Allow each tiny graph-paper square to represent 1 square inch (6.45 sq cm) of your grid. This allows you to make several reproductions of your grid on graph paper and to measure the hydrodynamic drag pattern of each hull.

2. Since this experiment requires the quick and accurate collection of data, it helps to have a friend assist you. Place hull A over the corks. At a signal, have your assistant pour a full pail of water into the top basin.

3. Observe through the grid as the water slides down the sluice and around the hull. What kind of drag pattern do you see? Does drag exist at both the front and back parts of the hull?

4. Use the grid to determine the exact length and width of the drag pattern, and reproduce the pattern on the graph paper.

5. Replace hull A with hull B and repeat the procedure, making notes as before.

Result: Each hull design displays a different pattern as water flows around it. The pattern appears at both the front of the hull (drag) and at the back end (wake).

Explanation

Designs with the least amount of resistance make the best hull shapes. Wedge-shaped models provide the least amount of drag at the front, and flat-shaped designs provide the smallest wake at the rear. So, for the most effective hull design, the best solution combines a wedge front with a flat rear. This ensures that water flows cleanly around the front, or bow of the boat, and that the separated streams are kept apart and do not collide at the rear, or stern of the boat.

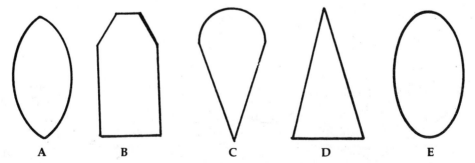

Hull Designs (part 2, step 1)

Display Tip

Reconstruct your sluice apparatus and exhibit it, labeling each part and its function. Display each of your hull designs (including original ones) and rate them on the success of their hydrodynamic designs. Display your test results in graph form.

Did You Know?

Good hydrodynamic design is not only important in the construction of ships but for any solid body that must stand or move through water. For instance, engineers must decide on the most effective shape for bridge pilings in order to keep the wear and tear of moving current from eroding the columns and weakening the bridge.

Can you think of other objects or natural sea or land creatures for which hydrodynamic shapes would be important?

Why Is a Kite Like a Sailboat?

You Will Need

- 2 wooden sticks 2 feet × ⅜ inch (60 cm × 0.9 cm), ⅛ inch (0.3 cm) thick (available in hobby shops)
- Kite string
- Wrapping tissue
- Scissors
- Rubber cement

This project demonstrates how the principle of *dynamic lift* applies to both kite and sailboat, with important differences.

Procedure

1. Place the sticks together in an X shape, and connect them by winding string around the joint in a crisscross pattern. Tie the string into a knot. You can also use a little rubber cement for extra strength.

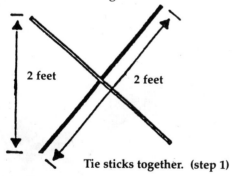

Tie sticks together. (step 1)

Here's a close-up view of sticks tied together in an X shape. (step 1)

2. Have an adult take the scissors and make a small groove near the end of each stick. This groove will help you wrap string around the sticks.

3. Tie the string to each stick at the notch with a single loop knot. Pull the string tight between the ends of the sticks as you tie the knots. The distance from one end of one stick to the next should be about 17 inches (42.5 cm).

Note groove in stick with string pulled tight. (steps 2–3)

4. Cut the tissue in a square so that it extends about 2 inches (5 cm) beyond the area enclosed by the string on all sides. Apply a little rubber cement, and fold the edges over the string all around the kite.

5. Turn the kite over so that the sticks lie underneath. Cut four 32-inch (80-cm) lengths of string, and attach the end of each string to a corner of the kite. Bring the strings together so that they form a kind of pyramid, and tie the long kite string to the point at which all the strings meet.

Tie strings together above kite. (step 5)

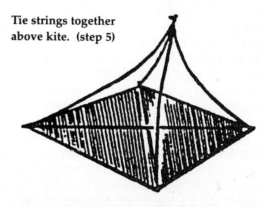

6. Make a tail for your kite by cutting an old sheet into 12-inch (30-cm) strips and tying three or four strips together. Attach the

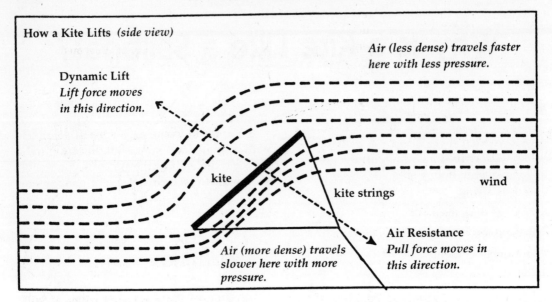

How a Kite Lifts (side view)

Dynamic Lift
Lift force moves in this direction.

Air (less dense) travels faster here with less pressure.

kite

kite strings

wind

Air (more dense) travels slower here with more pressure.

Air Resistance
Pull force moves in this direction.

tail to a corner of your kite. You may need to adjust the length of your tail as you begin to fly your kite.

7. Take your kite outside into a good wind, and gradually unwind the string. If you don't have sufficient wind, run with the kite.

Result: The kite rises at the end of its string. The stronger the wind or the faster you run, the higher the kite climbs.

Explanation
The climb of the kite can be explained by two forces: *air resistance* and *dynamic lift*. Air resistance makes it possible for any flat surface to glide in the air. But for a surface to actually lift, it must either become an airfoil or behave in a way that imitates an airfoil. *Airfoil* means "wing." A wing's design forces air to move over the top surface more rapidly than it moves over the bottom surface. Faster-moving air means that less pressure pushes down on the wing, and that the slower-moving air below can push the wing upward.

But since a kite really isn't a wing, what makes it lift? The answer is that the kite *behaves* like a wing. It tilts forward into the wind and is held in position by the string. This forces some air to spill over the top of the kite and some air to flow downward along its surface. The air streams meet at the same time behind the kite. Air flowing over the top has a greater distance to travel in the same time; that's why it flows faster.

Like an airfoil, faster-flowing air means air with less density and lower pressure. This means that air pressure is greater in front of the kite than behind it and that the kite moves upward and backward at an angle. At the same time, you pull the kite in the opposite direction, so that opposite forces balance each other and the kite stays nearly steady in the air.

The same principle of dynamic lift applies to a sailboat, but with some important differences. For one thing, a sailboat moves as the wind pushes against it, whereas the wing of an airplane moves *through* the air. But both wing and sail use a curved surface, or *airfoil*, to make them work.

How Is a Kite Like a Sailboat?

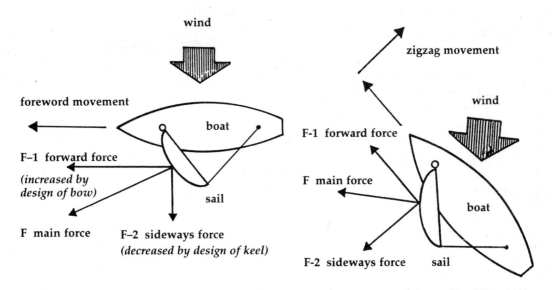

(A) Direction of Wind and Position of Sail for Forward Movement *(overhead view)*

(B) Direction of Wind and Position of Sail for Zigzag Movement (Tacking)

You've probably noticed how a sail billows out as the wind moves across it, creating a curved surface much like that of a wing. But on the sailboat, the "wing" is mounted vertically. As wind blows across the boat, it produces low pressure on the front, or billowing, part of the sail.

If you study the diagram (A) above, you can see how this low pressure creates a main force labeled "F." Part of this force (F-1) moves the boat forward through the water, while another part (F-2) tries to move it sideways. To reduce this sideways motion, some boats have a deep fin, or *keel*, extending down into the water from the underside of the boat.

If the wind comes from directly behind the boat, the sail simply swings out to the side so that the wind can blow directly against it. But if the boat tries to sail *against* the wind, the dynamics of wind against sail become a little more complicated.

The next diagram (B) shows how a sailboat can indeed sail against the wind, but only if it follows a zigzag pattern. This skillful sailing technique, called *tacking*, requires a well-designed boat with a deep keel. The angle of the zigzag depends on the sleekness of the boat. For racing boats, the angle may be as little as 40°, while more ordinary boats will tack at 50° or even 60°.

Display Tip

Display your kite. Use diagrams to explain how a kite is like a wing and how a wing is like a sail. Reproduce the diagrams showing the various kinds of wind forces against a sail and how tacking works.

Did You Know?

Kites were the first flying machines and were often put to practical use. Over a thousand years ago in Imperial China, kites were used as weapons. Huge kites were flown behind enemy lines to release streamers of burning phosphorus over enemy troops. Jumping ahead a few centuries, the Wright brothers' first plane was really a huge box kite, and when the great suspension bridge over Niagara Falls was begun, the first cable was carried across the falls by a kite.

Whirligigs & Parachutes

You Will Need

- 3 scraps of lightweight cloth
- Construction paper
- Fine string or thread
- Paper clips
- Scissors
- Ruler
- Pencil

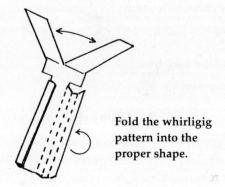

Fold the whirligig pattern into the proper shape.

Dynamic lift and air resistance are the two most important forces needed for lifting something and keeping it aloft. You can demonstrate both with a paper helicopter, or whirligig, and several cloth parachutes.

Part 1 Whirligigs

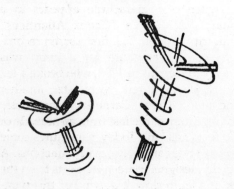

Procedure

1. Cut a large and small whirligig by following the pattern below.

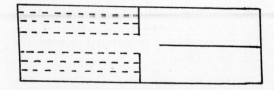

Whirligig Pattern (part 1, step 1)

2. Hold the smaller whirligig like you would a dart, and toss it into the air.
3. Observe the whirligig as it descends.
4. Repeat this procedure for the second whirligig, noting any differences in flight.

Result: After being thrown, each whirligig turns so that the shaft points down. As the whirligig begins to fall, the rotor starts to turn and slows its descent.

Part 2 Parachutes

Procedure

1. Cut square, triangular, and circular parachutes from the scraps of cloth.
2. Use a sharp pencil to make holes in the corners of the square and triangular parachutes; then make four equally spaced holes along the edge of the circular parachute.
3. Tie a separate piece of string through each hole in the cloth, and knot the strings together below each parachute.
4. Fasten five paper clips (or another small object, like a washer) to each knot for weight.
5. One by one, crumple up the parachutes and toss them high in the air. Observe which chute appears the most effective at slowing descent.

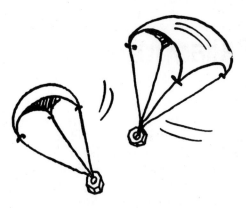

Parachutes with Small Weights

Result: The square parachute falls slowly and steadily, the triangular parachute falls more quickly, and the circular chute falls in an erratic, unstable pattern.

Explanation

Although whirligigs, like parachutes, demonstrate the principle of air resistance, only the rotors of the whirligig can create upward lift. Just as in a wing, air moving at different speeds above and below the tilted plane of the whirligig rotors creates a difference in air pressure. The slower-moving, high-pressure air below the rotors provides upward thrust. But in this case, the upward thrust doesn't lift the whirligig and only slows its descent.

Just as an inflated life belt holds a person afloat in the water, a parachute's canopy utilizes air resistance alone to slow its descent. It might seem like the best design for a parachute is the one that resists air the most. But in fact, a parachute canopy is a dynamic shape that must both resist fall and allow for it. This means that aerodynamic considerations come into play in successful parachute design.

Between your three parachute models, the square parachute has the best balance between air resistance and aerodynamics. As the square's four corners pull downward, four vents open up along the sides of the square allowing air to escape. This stabilizes the parachute. The triangular design also provides a stable canopy, but the smaller ratio of area to perimeter creates a parachute with less air resistance, making it fall too fast. A flat, circular parachute sacrifices stability for increased resistance, and this design will rock back and forth as captured air beneath the canopy escapes wherever it can.

Real-life circular parachutes aren't flat at all, but domed and vented for stability.

Display Tip

Have a friend photograph each whirligig and parachute test flight. Time the descent of each model and compare your results. Display the models and provide live demonstrations when someone requests them. Research and display photographs of different whirligig designs.

Did You Know?

Although Chinese acrobats used parachute-like devices as early as 1306, the first surviving sketch of a parachute appears in a Leonardo da Vinci's "Codex Atlanticus" manuscript of 1490. The first person recorded to use a parachute for a jump was François Blanchard in 1793, who broke a leg when landing. A few years later, another Frenchman, André Garnerin, successfully made five jumps, the last from an altitude of 8,000 feet (2,400 m). Today, parachutes come in many styles and have many functions. A specially designed parachute ejects from the back of a jet to slow it down. For this parachute, ribbons of fabric alternate with open net, replacing the solid canopy. The design allows air to pass through so that the chute can open at high speeds without tearing apart.

Boat-of-Clay

You Will Need

- 2 aquariums, same size
- 1 pound (about 450 g) of nonhardening modeling clay
- Small scale
- Blue food coloring (optional)

Did you ever wonder how a 200-ton (181-metric-ton) ocean liner stays above the water? By using modeling and two aquariums, this project demonstrates the principle of *buoyancy*.

Procedure

1. Fill each aquarium three-quarters full of water.
2. Mix just enough blue food coloring with the water to give it a bluish tint.
3. Divide the pound (450 g) of modeling clay in half, weighing each half to make sure you have two equal parts.
4. Mold half of the clay into a ball and the other half into a wide, hollow boat with tall sides.
5. Place the ball in the first aquarium, and place the boat in the second aquarium.

Result: The clay ball sinks to the bottom of the aquarium while the clay boat floats. The water level is higher in the aquarium containing the boat.

Explanation

Materials denser than water, such as metal or the modeling clay of this experiment, sink because they weigh more than the upward thrust of the water they push aside. But dense substances can be made to float by reshaping them to *increase* the amount of water they push aside. The greater the amount of water pushed aside, the stronger the upward thrust of the water—and so the boat floats.

Display Tip

Invite onlookers and judges to model the clay into their own boat shapes. Some boat shapes float better than others, since everything depends on good design.

Did You Know?

To solve the overcrowding problem in the city of Tokyo, Japanese engineers have designed a gigantic raft to float in Tokyo Bay. This raft would consist of super-compressed garbage molded into interlocking logs. The raft would be so strong that buildings and roads could be constructed on it, extending the city out over the water. Such a huge displacement of water might require dams along the shore that would provide the rafts with hydroelectric power. It's all just another reason to appreciate the wonders of science and good engineering.

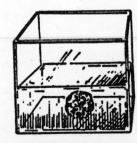

Aquarium #1 with Ball

Aquarium #2 with Boat

Metric Equivalents

Capacity (Liquid & Dry Measures)
1 milliliter = 0.2 teaspoon = 0.07 tablespoon = 0.034 fluid ounce = 0.004 cup
1 teaspoon = 100 drops = 5 milliliters = ⅓ tablespoon
1 tablespoon = 3 teaspoons = ½ fluid ounce = 15 milliliters
1 fluid ounce = 2 tablespoons = 30 milliliters = 0.03 liter
1 cup = 16 tablespoons = 8 fluid ounces = 240 milliliters = 0.24 liter
1 pint = 2 cups = 480 milliliters = 0.47 liter
1 quart = 4 cups = 32 fluid ounces = 960 milliliters = 0.95 liter
1 liter = 1,000 milliliters = 61.02 cubic inches = 34 fluid ounces = 4.2 cups = 2.1 pints = 1.06 quart (liquid) = 0.908 quart (dry) = 0.26 gallon
1 gallon = 4 quarts = 128 fluid ounces = 3.8 liters

Weight (Avoirdupois)
1 gram = 0.035 ounce = 1,000 milligrams = 0.002 pound
1 ounce = 28 grams = 437.5 grains = 0.06 pound
100 grams = 3½ ounces
1 pound = 16 ounces = 454 grams = 0.45 kilogram = 7,000 grains
1 kilogram = 2.2 pounds = 1,000 grams
1 ton = 2,000 pounds = 0.9 metric ton
1 metric ton = 1,000 kilograms = 1.1 tons

Length & Distance
1 millimeter = 0.039 inch
1 centimeter = 10 millimeters = 0.39 inch = 0.03 foot
1 inch = 25 millimeters = 2.54 centimeters = 0.025 meter
1 foot = 12 inches = 30 centimeters = 0.3 meter
1 yard = 3 feet = 36 inches = 90 centimeters = 0.9 meter
1 meter = 100 centimeters = 1,000 millimeters = 39.37 inches = 3.28 feet = 1.09 yards or 1 yard + 3.4 inches
1 rod = 5.5 yards = 16.5 feet = 5.03 meters
1 kilometer = 1,000 meters = 0.6 mile
1 mile = 1,609.3 meters = 1.6 kilometers

Area
1 square centimeter = 0.15 square inch
1 square inch = 6.45 square centimeters
1 square foot = 0.09 square meter
1 square yard = 0.83 square meter
1 square meter = 10.76 square feet = 1.19 square yards
1 square rod = 30.25 square yards = 0.0062 acre = 25 square meters
1 acre = 43,560 square feet = 4,840 square yards = 160 square rods = 0.4 hectare = 4,047 square meters
1 hectare = 2.47 acres
1 square kilometer = 0.38 square mile
1 square mile = 640 acres = 2.59 square kilometers

Volume
1 cubic centimeter = 1,000 cubic millimeters = 0.06 cubic inch
1 cubic inch = 16.38 cubic centimeters
1 cubic foot = 0.028 cubic meter = 1728 cubic inches = 0.037 cubic yard
1 cubic yard = 27 cubic feet = 0.76 cubic meter
1 cubic meter = 1,000,000 cubic centimeters = 35.3 cubic foot = 1.3 cubic yards

Temperature
To convert **Fahrenheit** to **Centigrade** degrees, use this formula:

$$5/9 \, (°F - 32) = °C$$

To convert **Centigrade** to **Fahrenheit** degrees, use this formula:

$$9/5 \, °C + 32 = °F$$

Centigrade is also called Celsius.

Index

A
aa, 70
abrasion, 148
absorption, 63, 64
abstract, 7, 8
acid rain, 88
acids, 68, 113
  corrosive to plants, 171
  as electrolytes, 113
  as preservatives, 75–76
acrobats, 199
active metal (anode), 114–115, 122–123, 155, 157
aerodynamics, 191, 195–197, 198–199
ailerons, 186–188
air
  carbon dioxide in, 174
  convection current, 74
  crosswinds, 74
  density, 196–197
  dynamic lift, 195–197, 198–199
  flow, 196–197
  friction, 191
  high-pressure, 199
  humidity, 66, 85
  low-pressure, 196–197
  motion in flight, 188
  pressure, 196, 199
  resistance, 196–197, 198–199
  rotor-vortex microbursts, 74
  in tornado, 73–74
  warm/cool fronts, 74
airfoil, 196
airplane; see also flight
  aileron, 186–188
  ascent (climbing), 188
  banking (turning), 188
  box kite as first, 197
  control stick, 188
  descent, 188
  elevators, 187–188
  fuselage, 186–187
  glider, 185–188
  helicopter flight, 198
  lifting, 185–188
  low-flying, 74
  magnetometers, 35
  microbursts of air, 74
  nose, 188
  rudder, 188
  tail, 185–188, 195–197
  turning, 185–188
  wind test for, 187–188
  wing (airfoil), 185–188, 196–197, 198–199
  Wright brothers, 197
albumin in, 101
alcohol, 61, 71
algae, 163–164
alkaline-acid reaction, 68
alloy, metal, 155
alternating current, 126
altitude, 14, 183, 199
aluminum, 116–117, 122–123, 131, 155, 157
ambergris, 62
angle
  altitude, 14
  escape velocity, 143–144
  of inclination, 143–144
  of hillside objects, 150–151
  and parallax principle, 27–33, 182–184
  of tree lean, 150–151
  of triangle, 28, 29, 32–33, 183
anchor threads, spiderweb, 55–56
angular diameter, 29
angular relationship, sun and sky, 17
animals, 57, 77, 86, 170
anode, 115; see also metal(s), active
ant
  antennae, 52
  architecture, 43–44
  army, 44
  cemetery, 44
  chemoreceptors, 52
  colony, 43–44
  color, favorite, 58–59
  collecting, 43
  gardens, 44
  hill, 51
  larvae, 44
  mandible, 53
  moat, 43
  mounds, 44
  nurseries, 44
  pheromones, 52
  queen, 43–44, 52
  social order, 44
  trails, 44, 51–52
  tunnels, 44
  warrior, 52
antennae, insect, 52
apple
  juice, 90–91
  painted, 96
  ripening, 96, 167
  starch in, 96
  sugar and, 94–95, 96
  tree grafts, 170
  varieties, 170
applesauce, 94–95
appliances and electricity use, 118–119
aquarium, 53–54, 189–190, 200
Araneae, 55
archway bridge, 135–136
area, circle, 29
area, ratio to perimeter in parachute, 199
Aristarchus of Samos, 27
Aristotle, 161
army ants, 44
aromatic oils, 61–62
artichoke, 97
asparagus, 107–108
aspartamine, 97
astronomers/astronomy, 14, 17, 18, 20, 22, 24, 27, 30–31, 183
atmosphere, Earth, 16–17
  electrical charges, 116
atmospheric pressure
  hummingbird feeder, 49–50
  liquid flow and, 49–50, 141–142
  measuring, 137–138
  plunger test, 137–138
  self-filling water dish for pets, 141–142
  vacuum and, 137–138
attraction, electrical/magnetic, 67, 116–117, 120–121, 124, 129, 130, 131, 146–147, 154–155, 156–157
avocado, 92–93, 167–168, 171–172
avocado dip recipe, 92
axle, gear/wheel, 160

B
backboard, 7, 8
bacteria
  eat iron, 108
  food preservatives and, 75–76
  growth, 75
  magnetite formation, 108
  plastics breakdown, 78
  in rocks, 108
bacon, fat test, 92–93
baffle, wind, 185–188
baffle test for glider, 185–188
baked goods, 106
baking soda, 68, 79, 113, 122
  as electrolyte, 113, 122
baking-soda solution, 79
balance, Earth-Moon, 145
balloon and static electricity, 121, 129
ballpoint pen metals, 155
banana, 99–100, 167–168
  food for insects, 51–52, 53–54
  ripening, 167–168
  trees, 79
barbs of feathers, 46
barbs/hooks of seeds, 165
barbicels of feathers, 46
barium, 35
bark, tree, 166
barycenter, Earth-Moon, 145
battery
ambergris, 62
car, 132
D-cell, 132
poles, 132–133
positive and negative terminals, 132–133
with potato polarity indicator, 132–133
six-volt (6-volt), 112–113, 114–115, 127–128
beavers, 46
bean(s), 100
bee, 58–59
Berlese funnel, 45
berries, 170
billiard balls, 77
binoculars, 30
biodegradable materials, 77–78
Bird Feathers & Bug Boxes, 42–59
bird feeder (hummingbird), 49, 57
bird
  craving fat, 57
  -feeders, 49, 57
  pudding, 57
  in winter, 57
bitter taste, 98
Blanchard, Francois, 199
blue in chemical reaction, 90–91
blue light, 16–17
blue sky, 14–15, 16–17
  cyanometer, 14–17
  at horizon, 14–15, 16
  pollution and, 16–17
  time of day and, 16
  unique, 17
boat; see also ship
  clay, 200
  hull design, 191–194
  hydrodynamics, 191–194
  sailboat movement, 195–197
bodies, orbiting, 145
body, human, 38, 63, 65, 90–110, 137
boiling point (water), 85
bone minerals, 110
botanists, 169–170
box periscope, 38–39
branched venation, 79–80
brass, 115, 123, 157
bread, 99–100, 106
bridge construction
  archway, 135–136
  girder, 135–136
  materials, 135–136
  pilings, 194
  plank, 135
  porous structure, 67
  structure, strength, and sag, 135–136
  triangular-truss, 136
  types, 135–136
brittle metal, 115
bronze, 123
broth, spoiling, 75
brown paper test for fats, 92–93
bubble-gum plant graft, 169–170
bugs, 45, 57, 58–59
building structure, 67
bulbs, plant, 171
bumper, floating, 158
buoyancy, 46, 200
butter, fat test, 92–93
butterfly, 53–54, 56, 58–59
  color, favorite, 58–59
  elephant and, 53–54
  feeding, 53–54
  nectar, 54
  proboscis, 53–54
buttons, metal, 157
button-making, plastic, 77–78

C
calcite (calcium carbonate), 63, 87
calcium, 110
calculator use, 32–33, 174
caldera, 69
calories, fat, 93
calories, postage stamp, 105
caloric energy and swimming, 93
cane sugar, 95
canopy, parachute, 199
capillary action, 67
car
  battery, 132
  jumper cable, 132
  shock absorbers, 158–159
  speedometer, 131
carbohydrates, 110
carbon dioxide, 63, 68, 167, 174
cardboard stand, 177, 189
cardinal phases, moon, 19, 20
carrots, 171–172
casein, 78
Cathedral of St. Isaacs, 140
cathode (less active metal), 115, 122–123, 155, 157
caution, 6, 43, 55, 58–59, 76, 86, 95, 99, 100, 101, 165, 173, 178, 180; see also specific project directions
cave "icicles," 66–67
CD player, 131
celery stalk, circulation/dyeing, 166
cell
  growth and repair, 110
  largest living, 64
  membranes, 63, 64, 65
  metabolism, 63, 64, 65, 180
  nutrient intake, 180
  red blood, 107
  wall demonstration, 63, 64, 180
  wastes, 180
cellophane, charged,

Index

124
centrifugal force, 152–153
cereal, iron content, 109
cereal (puffed wheat), charged, 120–121
charges, electrical, 116–117, 121, 124, 129, 130
chart
 atmospheric pressure, 138
 angular altitude, 14–15
 bug's favorite color, 58–59
 clinograph measurements, 183–184
 creeping soil (tree lean on), 150–151
 cyanometer, 14–15, 16–17
 depth, ocean, 40–41
 echo-location, 40–41
 electric consumption, 119
 ferromagnetic flowchart, 156–157
 insect color preference, 58–59
 leaf grid, surface area, 174–175
 of metals and magnetic attraction, 154–155
 rock tumbling, 148–149
 surface area, atmospheric pressure, 137–138
 surface area of leaf, 174–175
 topography of ocean floor, 40–41
 tree lean, 150–151
chemistry, 52, 61–62, 85–86, 96, 99–100, 114–115, 132, 180
chemical; *see also specific chemical or test*
 activity, 114–115, 122–123
 bonding, 91
 energy, 175
 indicator, 109
 messages (pheromones), 52
 reaction, 85-86, 96, 100, 101–102, 180
chemically active metals, 114–115, 122–123
chemoreceptors, 52
chicken, raw and cooked, 101–102
Chinese acrobats and parachutes, 199
Chinese barn-fire, 101
Chinese kites as weapons, 197
chitin, 78
chitisand, 78
chlorophyll, 163
circle
 area, 29
 boat hull, 191–194
 circumference, 29
 leaf shape, 176–177
 parachute, 198–199
 circuit, electrical, 112–113, 127–128
 breaker, 113
circulatory system of plant, 166

cities, ancient, 88
cities, overcrowding, 200
clays to strengthen paper, 82
climate, humid, 66
climate, dry, 172
clinograph, 182–184
cloudiness, 75–76
clover, 59
coagulation, 101
cobalt, 155
cobwebs, 55
Codex Atlanticus, 199
cola, 94–95
collecting data, 5
cologne (waters), 61
color
 altitude, sky, 14–15
 blue, 48, 58–59
 blue light, 16–17
 blueness of sky, 14–17
 bluish black (iodine/starch tests), 180
 celery demonstration, 166
 cyanometer, 14–15, 16–17
 food, 166
 green, 58–59
 green pigment in plants, 163
 high-spectrum, 48
 insect preference, 48, 58–59
 perception by insects, 48, 58–59
 pollution, sky, 16, 17
 purple, 48, 58–59
 red, 58–59
 red light, 17
 red sunlight, 16
 and ripeness, 167–168
 sky, 14–17
 ultra-high spectrum, 48
 ultraviolet, 48, 59
 violet, 48, 58
 wavelengths, 59
 white sunlight, 16
 yellow-jacket bees and yellow, 59
comb electrical charge, 118–119, 120–121
combining liquids, 83–84
combustible properties, 77
compass, 11, 12–13, 14, 18, 125–126, 127
 cardinal points of, 14
 as galvanometer, 125–126
 navigator's, 14
 response to electrical current, 125–126
conductivity, 114
conductor, electrical, 113, 114, 125
acids/ acid salts, 113, 132
basic salts, 113
controlling current, 127–128
in electro- or copper-plating, 114–115
liquids, 113
neutral salts, 113
poor, 127
rheostat, 127–128
constellation patterns,

21, 22
conservation
 compressed garbage, 200
 of electricity, 118–119
 of fuel, 144
 overcrowding and land, 200
 recycling paper, 81–82
 space station, 143–144
continents, 145
continuous loop of electricity, 112
controlling current, 127–128
convection current, 74
cookie, fat test, 92–93
cooled water expands, 26
copper, 114–115, 122–123
copperplating, 114–115, 122–123
copper-sulfate solution, 99–100
copper wire, 112, 125–126, 127–128
copper-wire (induction) coil, 125–126
corn, starch content, 104, 105
corn, sugar content, 94–95
cornstarch (solution test), 71, 90–91, 180
corn syrup, 63, 95
corrugations, material strength, 135–136
cotton-milling, 77
cow magnet, 37
cranberry juice, 107
crater, 19, 69, 145
Crater Lake, 69
cream of tartar as electrolyte, 113
cream, fat test, 92–93
creamy plastic, 77–78
Creative Concoctions, 60–88
creeping soil, 150–151
crescent moon, 19
crosswinds, 74
crust, Earth's, 71–72
cultivation of plants, techniques, 171–172
culture, 75
curds, 77
current, electrical, 112–113, 127–128
cuttlefish, 62
cutworms, 45
cyan, 14
cyanometer, 14–16
D
dams, 200
da Vinci, Leonardo, 199
debris, 74
decomposition of plastics, 77–78
decrescent moon, 19
degree/ angle, 29–30, 150–151
dehydration, 65
density, air, 196–197
depth sounding, 189–190
design
 airplane glider, 186–188
 baffle, wind, 185–186
 bridge pilings, 194

boat hull, 191–194
city, floating, 200
 for flight, 185–188, 195–197, 198–199
 flotation, 191–194, 200
 leaf (shape), 176–177
 parachute, 198–199
 or pattern, plant, 173, 176–179
 sluice, 191–192
 strealined ship, 191
 water runoff, 176–177
detergent, 46, 70, 83, 163–164
deuterium, 85–86
dextrin, 103, 105, 106
dextrose, 95
dicotyledons, 79, 80
dipping compass, 12–13
disease and heavy water, 86
disease and vaccination, 170
dispersion, 147
displacement, water, 200
dissolving metal, 115
distance, measuring, 27–29, 182–183
 apparent movement, 183
 clinograph, 182–184
 between Earth, Moon, planets, and stars, 30–31, 183
 and height calculation, 32–33
 kite-string, 29
 line of position, 28, 31
 to Moon, formula, 31
 pacing off, 184
 parallax effect, 183
 parallax principle, 27–29
 parallax shift, 30–31
 between stars, 30–31
 triangulation instrument, 27–29
 triangulation, 27–29, 30–31, 32–33, 183
dogs, vitamin C, 90
drag (pattern), 193
"drainpipes," leaf, 176–177
drop-by-drop measurement, 90
droplet accumulation, 67
drought, 172
drying food, 75
ducks, 46
dust-rag, wool, 46
dynamic lift, 195–199
E
Earth
 balance (barycenter) with Moon, 145
 bulge, 152–153
 crust, 71–72
 diameter, 31
 distance to stars, 21
 distance to Moon, 29, 30–31
 gravity, 140, 143–144, 152–153
 geographic poles of, 11, 12–13
 gravitational pull, 140, 152–153
 interior, 71–72
 layers, 71–72

magnetic field of, 10–11, 12–13
magnetometers, 35
mineral deposits, 35
soil composition, 35
magnetic poles of, 10–11, 12–13
mantle, 71–72
pendulum swing and, 139–140
rare metals, 34–35
rotation, 12–13, 31, 139–140, 152–153
solid surface, 153
space station and, elevator, 144
speed, 152
tides, 153
zone of partial melting, 71–72
earwigs, 45
echo-location (-sounding) on ships, 40–41
formula, 40
eddy-current motor, 131
eddy-flow principle, 131
edible iron, 109
egg, 63, 64, 84, 99–100
albumin in, 101
cell, largest living, 64
chemical composition, 63
in mayonnaise, 84
protein-rich, 100
shell of, 63
sugar-shriveled, 64
yolk as emulsifier, 84
white, 99–100
egrets, 46
electrical circuit
closed, with magnets, 112–113
controlling current, 127–128
copperplating, 114–115
spring and, 112–113
electrical charges
attraction/ repulsion, 116–117, 124, 129, 130
flash dancers, 124
like (charges) repel, 116, 129
losing, 121
negative/ positive, 116–117, 124, 129, 130
opposite (charges) attract, 116–117, 129
particles, 129, 130
static electricity, 124, 129, 130
electrical conductor, 113, 125–126, 127–128
electrical current, 112–113, 125–126, 127–128, 132–133
electrical induction, 125–126
electrically neutral, 120
electricity
alternating current, 126
appliances, 119
consumption, 119
cost of, 118–119
conservation, 118–119
dams and hydroelectric power, 200
friction, 116–117, 129, 130
galvanometer, 125–126
generators, 125–126

Index

induction coil, 125–126
magnetism, 125–126, 131
meter reading, 118–119
motor, 131
projects, 111–133
solar panel, 174–175
static, 116, 120–121, 124, 129, 130
electrolyte, 113, 114, 122–123, 132–133
  acids as, 113, 132–133
  liquids as, 113, 132–133
  potato, 132–133
  salts as, 113
electromagnetism, 113, 131, 140, 144
Earth's rotation, 140
elevator, to space station, 144
  motor, 131
  pendulum, 140
  spring, 113
electrons
  boxed, 120–121
  charged, 120–121, 129, 130, 132–133
  in electroplating, 114–115, 122–123
  heat speeding, 122
  neutral, 121
  puffed wheat, 120–121
  rust, patina, or tarnish and, 122–123
electroplating, 114–115, 122, 155, 157
  anode (active metal), 115, 122–123
  cathode (less active metal), 115, 122–123
  copper(plating), 114–115, 122
  electric current, 114
  electron loss, 114–115, 122–123
  and ferromagnetic metals, 155, 157
  and removing silver tarnish, 122–123
electroscope, 116–117
electrostatic flower, 129
elephant, 53–54
elevator to satellite, 144
elevators, airplane, 185–188
elm trees, 79
elodea, 163–164
embryo, plant, 173
emulsifiers, 83, 84
energy
  changing, 174–175
  chemical, 175
  efficiency in boat hull design, 191–194
  electrical from sun, 174–175
  fats and oils, 92–93
  iron and, 110
  oxygen, 110
  starch, 104, 174–175
  transforming, 174–175
engine, rocket, 143–144
engineering, genetic, 170
engineers, 131, 161, 191, 194, 200
Epsom salt, 66–67
Equator
  bulge at, 152
  Earth's rotation, 152–153
  magnetic force at, 12
  and swinging pendulum, 139–140
erosion, soil, 25, 150–151
escape velocity, 143–144
ethylene, 167–168
eutrophication, 163–164
evaporation, 61, 67, 92
evergreens, 79
evolution, 137, 166
expansion of ice, 26
experiment, basics, 6
explosion, marble powder, 88
Excellent Electrics, 111–133

F

Faraday, Michael, 125–126
fat(s)
  animals, 57, 93
  energy, 57, 92–93
  in foods, 92–93, 110
  heat and, 57, 92, 93, 101
  layer beneath skin, 93
  melting point, 92
  non- and low-, 93, 95
  and oils, 92–93, 110
  as protection, 93
  sea creatures and, 93
  as solids, 92
  swimming and, 93
  temperature and, 92
  testing, 92–93
  warmth, 93
feather, 46
feeder, bird, 49, 57
fence post, lean, 151
ferrites, 35
ferromagnetic flowchart, 156–157
ferromagnetism, 36, 155, 156–157
fertilizer, 78, 163–164, 172
fertilizing flowers, insects, 58–59
fiber-optic tube, 38
fiber, paper, 81–82, 92
field guide, 165, 178
filtration, magnetic, 36–37
Find Out about Food, 89–110
fingernails, 78
fire and marble, 88
fish, preserving, 75–76
fixative, 62
flaps, airplane wing and tail, 185–188
flash dancers, 124
flashlight bulb (lamp tester), 127–128
flies, 58–59
flight, 185–188, 191, 195–197, 198–199
airplane glider, 185–188
helicopter, 198
kite, 195–197
parachute, 198–199
warfare and kite, 197
whirligig, 198–199
flotation, city, 200
flotation, life belt, 199
floating feather, 46
flowers, 45, 58–59, 79
flower, electrostatic, 129
Fly, Float & Sink, 181–200
food
  bacteria growth, 75–76
  bitter taste, 98
  bones, 110
  for cell, 180
  cooking, 101–102
  coloring, 46, 49, 68, 83, 100, 166
  combining for protein, 99–100
  enhancements, 97
  fat test, 92–93
  flavor, 61, 97–98, 167–168
  fortified, 109
  heat and, 92, 101–102, 106
  instant, 104
  iron test, 107–109
  liquid, 75
  manufacture/transport in plants, 166
  metals in, 36–37
  minerals in, 110
  oil in, 92–93
  pickled, 75
  preserving, 75–76
  processing, 36–37, 94
  projects, 89–110
  protein test, 99–100
  ripening fruits and vegetables, 167–168
  salt(y), 75–76, 97–98
  solid, 109
  sour taste, 98
  spoiling, 75–76
  starch tests, 103–105
  sugar test, 94–95
  sweet, 75, 94–95, 97–98
  taste, 97–98
  toasting, 106
  vitamin C test, 90–91
force
  air on sail, 196–197
  centrifugal, 152–153
  electromagnetic, 131
  gravity, 67, 140, 143–144, 152–153, 158, 159
  magnetic lines of, 158
  momentum (sideways swing), 140
  repulsion, 158
  third, 140
formula
  angular diameter, 29
  calculator use of, 32–33
  circles and, 29
  distance to Moon, 31
  distance to stars, 27–29, 30–31
  echo equation, 40
  ocean depth, 40, 41
  parallax principle, 29
  sound through water, 40
  triangles and, 32–33
  trigonometric for calculating distance, 27
  trigonometric for calculating height, 32–33
Foucault, Jean, 139
Foucault's pendulum, 139–140
freezing landscape, 26
friction, air, 191
friction and electricity, 116–117, 124
fructose, 95
fruit
  gas ripening, 167–168
  greenhouse, 167–168
  how to tell from vegetables, 170
  hybrids, 170
  iron content, 107–108
  protein, 99–100
  ripening, 167–168
  seeds of, 170
fuel conservation, 144
fulcrum and lever, 161
full moon, 19
fungi, 170, 173, 178
funnel in depth indicator, 189–190
funnel of tornado, 73–74
fuselage, 186

G

galvanized iron, 85
galvanometer, 125–126
garbage, super-compressed, 200
gardening, 172
Garnerin, Andre, 199
gases, molten-lava, 69
gears, motion of, 119, 160–161
gelatin, 83, 101–102
generators, 125
genetic engineering, 170
geographic South Pole, 11
geology (study of Earth), 71–72, 151
geometry, 32–33, 183
German mayor's vacuum experiment, 138
germination, 165
gibbous moon, 19
gills, 173
girder bridge, 135–136
gladiolas, 79
glass, 71–72
glider, 185–188
glucose, 94–95
glucose strips, 94–95
"gluing," 122
gnat, 58–59
grafting plants, 169–170
grain(s), protein, 99–100
grain size, erosion, 25
grain, starch, 103
granite, 149
grape juice, 90–91, 107–108
grapefruit, 95
graphite, 127–128
grasses, 79, 172
grasshoppers, 53
gravestone, tilt, 151
graveyard marble, 72
gravity, 67, 140, 143–144, 152–153, 158, 159
  centrifugal force and, 152–153
  escape velocity and, 143–144
Great Plains, 41
Greeks, ancient, engineering text, 161
  parallax principle, 27
green patina, 123
green pigment, 163–164
green pepper juice, 90–91
green vs. ripe fruit, 167–168
greenhouse, 168
ground movement, 150–151
groundwater, polluted, 46
guacamole dip, 92
gum (bubble-) graft, 169–170

H

half-moon, 19
hamburger meat, 99–100
Hawaii, volcanoes, 70
healing, vitamin C, 90
heat, electric dryer, 128
heat speeding electrical interaction, 122
heavy water, 85–86
height, calculating, 32–33, 182–184
hematite, 35
hemoglobin, 107, 110
hill, movement, 150–151
hitchhiking seeds, 165
homemade
  paper, 81–82
  perfume, 61–62
  plastic, 77–78
honey lure, 58, 59
horizon, 14, 16
hormone, ripening, 167–168
horses and vacuum, 138
household appliances, electricity, 119
hull, boat, 191–194
human body
  atmospheric pressure on, 137
  cell growth and repair, 99, 107
  evolution, 137
  fiber-optic tube, 38
  nutrients, 63, 65, 90–110
hummingbird feeder, 49–50
hybrids, 170
hydrodynamics, 191–194
hydroelectric power, 200
hydrogen, 85–86
hydroponic garden, 171–172
hypothesis, scientific, 5–6, 8

I

ice cream, 84

Index

ice crystals
  splitting stone, 26
  in thundercloud, 116
ice, expansion, 26
"icicles," cave, 66–67
ilmenite, 35
immiscible liquids, 83–84
Imperial China, 197
inclination, angle of, 143–144
inclinometer, 12–13
induction, 126
induction coil, 125–126
inertia, 143
infection, preventing, 90
insects
  architecture, 43–44
  beneficial, 48
  Berlese funnel, 45
  color perception/preference, 48, 58–59
  leaf-eating, 178
  light-seeking, 47–48
  mandible, 53
  navigation, 48, 59
  nocturnal, 47–48
  parasites, soil, 45
  pheromones, 51–52
  proboscis, 53–54
  projects with, 43–45, 47–48, 51–54, 58–59
  sampling, 47–48
  soil-burrowing, 45
  trails, 51–52
  trap, nocturnal, 47–48
  visual acuity, 48
iodine test(s), 90–91, 96, 103, 105, 106, 180
  caution, 90, 180
  cell-wall semipermeability, 180
  molecule size, 180
  stain, 90, 180
  starch solution, 180
  titration, 90
iron
  alloy, 155
  bacteria eating to form magnetite, 108
  in cereal, 109
  combined metals, 157
  Earth's magnetic field and, 10–11, 35
  edible, 109
  ferrites and, 35
  flowchart (ferromagnetic), 156–157
  in foods, 109, 110
  hematite (iron ore), 35
  hemoglobin, role in producing, 107, 110
  isolating, 109
  magnet and, 131, 146–147, 154–155, 158–159
  magnetite and, 11, 35, 36, 108
  magnet poles, 158–159
  oxide (Fe_2O_3), 17, 85
  oxygen carrying, 110
  plating and, 157
  rusting, 122–123
  sand and, 36–37
  soil and, 34–35, 36–37
  testing, 107–108, 109
  yttrium-iron garnet, 35

J

Japanese scientists/engineers, 78, 200
jeweler, 148
journal keeping, 6
judges, 6–8; see also specific project
juice tests for iron content, 107–108, 109
juice tests for vitamin C, 90–91
jumper cable for car, 132

K

keel, boat, 197
ketchup (catsup), 94–95
key, copperplating, 114–115
key, metal composition, 114–115, 154–155
kidney beans, 99–100
kilowatt/kilowatt-hours, 118–119
kite
  dynamic lift, 195–197
  historical uses, 197
  sailboats and, 195–197
  -sighter, 182–184
  -string measurement, 29

L

lactose, 95
lambchop, raw and cooked, 101–102
lamp tester, 127–128
landfill, 25, 77–78, 82
land mass drift, 71–72
landslide, 151
lanolin, 46
largest living cell, 64
largest molecule, 103, 180
latitude, 12, 152
lava
  ejecting, 68, 69
  fluid or crumbly, 68–69, 70
  Hawaiian names, 70
  molten, 69
lean, objects on hill, 150–151
leaves
  arrangement of, 178
  branched venation, 179
  collecting, 178–179
  counting, 174–175
  dissolving, 79–80
  identifying, 178–179
  parallel venation, 179
  points, small, 177
  ribs, 178–179
  rubbings, 178–179
  shape of, 176–177
  skeletonized, 79–80
  solar panel and, 174–175
  stomata, 179
  as sunlight collectors, 174–175
  surface area of, 174–175, 176–177
  venation, 79, 178–179
  water runoff, 176–177
lemon, sugar content, 95
lemon juice, 96, 113, 132
  as electrolyte, 113, 132
lever and fulcrum, 161
life belt, flotation, 199
ligaments, 102
light
  bending, 22
  and biodegradable plastics, 78
  insect navigation, 48
  wavelengths, 16–17
  -years, 22
lightbulb, 118–119
  rheostat (controlling intensity), 127–128
lightning, 124
lignum, 82
lilies, 79–80
limestone, 87, 148–149
line of position (star movement), 28–31
line(s) of force, magnetic, 10, 131, 147, 158
liquid
  cooling, 26
  conductors of electricity, 113
  expansion, 26
  fat, 92
  flow of, 71
  glass, 71–72
  immiscible, 83–84
  internal friction, 71
  molten rock, 71–72
  sandstone, 148
  and solid, 26, 71–72
  super-cooled (inside Earth), 72
  viscosity and, 71–72
liquid fertilizer, 172
livestock blood, 77
livestock, metal fragments and, 36–37
living cave, 66
locking fibers, 82
London Science Museum, 140
lunar quarters, 18, 19–20
lye solution, 99–100
Lyme disease, 165

M

machine, gears, 160–161
Magdeburg, Germany, 138
magnesium, 35
magnet(s)
  bar, 12, 109, 125–126, 146–147, 154, 156, 158–159
  battery poles, 132–133
  electric circuit, 112–113
  electric motor and, 131
  in electricity generators, 125–126
  floating, 158–159
  eddy-flow principle and, 131
  horseshoe, 34, 36, 131, 147, 154, 156
  induction/conduction, 125–126
  poles, 10–13, 112–113, 132–133, 146–147, 158–159
  rotating, 131
  round, 147
  spring, 112–113, 158–159
  wire coil, 112
magnetic
  attraction/repulsion, 12–13, 113, 131, 146–147, 154–155, 158–159
  Earth's field, 10–11, 35
  field, 10–11, 35, 126, 131, 146–147
  filtration, 36–37
  force (lines), 10, 131, 147, 158
  particles, soil, 34, 35
  poles, 10–13, 112–113,
158–159
  spring, 158–159
  3-D field, 146–147
  magnetic poles, Earth's, 10–11, 12–13
    vs. geographic North and South Pole, 12–13
  magnetism, 12, 125–126, 131, 158
  magnetite, 11, 36, 108
  magnetometers, 35
  magnifying glass, 34, 165
  magnifying gradient, 53
  magpies, 57
  maltose, 95, 103
  mandibles, 53
  manganese, 35
  mantle, Earth's, 71–72
  maple leaves, 79
  maple syrup, viscosity, 71
  mapping, ocean, 40–41
  marble
    acid rain and, 88
    atmospheric pollutants and, 88
    erosion, 87
    explosion, powder, 88
    flows, 72
    heat, fire, and, 87–88
    quarry, 87
    rolling glass (escape velocity), 143–144
    test for, 87–88
    vinegar and, 87
  Marianas Trench, 41
  Mars, 17
  materials
    alternate, 5
    arch construction, 135
    concentration, 64
    corrugated, 136
    electricity flow, 147
    flow, 63, 64, 65
    magnetism flow, 147
    shape, 135–136
    space-station, 144
    strength, 135–136
  mechanical advantage, 161
  mechanical motion, 160–161
  timing, 161
  measurement, drop-by-drop, 90–91
  meat
    iron content, 107
    preserving, 75–76
    raw and cooked, 101–102
  melting point, 92
  melting, zone of partial, 71–72
  membrane, semipermeable, 63, 65
  mesocyclone, 74
  metal(s)
    active, 114–115, 122–123, 155, 157
    alloy (combined), 154–155
    attractive and nonattractive, 154–155
    brittle, 123
    changing, 122–123
    chemically active, 114–115, 122–123
    combining, 156–157
    electrically charged,
114–115, 116–117
  electroplating, 114–115, 122–123, 157
  ferromagnetic, 155
  ferromagnetic flowchart, 156–157
  magnetic attraction of, 154–155
  nonmagnetic, 154–155
  patina, 123
  plating, 155, 157
  rare-earth, 34–35
  rusty, 85–86, 122–123
  tarnish, 122–123
  test for ferromagnetic, 154–155
meter, electric, 118–119
metric equivalents, 8, 201
Michigan, trees, 179
microbursts, rotor-vortex, 74
milk, 77–78, 93, 94–95
milk paint, 77
milk-product plastic, 77–78
mineral(s), 110
  deposits, 66–67
  food containing, 107
  oil, 146–147
  stalactites and stalagmites, 66–67
mirrors, 38
mixture, 147
mnemonic, 67
modeling clay, 143, 145, 200
models, constructing, 5
moisture in air, 85
molecules, 63, 64, 65, 71–72, 78, 91, 103–104, 180
  chemical bonding, 91
  density, 83–84
  glass, long chains, 71–72
  internal friction, 71
  plasticlike, 78
  rubbery, 78
  size, 180
  starch, 103–104
molten (sea; rock), 71–72
momentum, 143
monocotyledons, 79–80
Moon, 48
  barycenter (balance) with Earth, 145
  box, 18–20
  cardinal phases, 19, 20
  crescent, 19
  decrescent, 19
  gibbous, 19
  half-, 19, 20
  new, 19, 20
  quarters, 19, 20
  waning/waxing, 19
  craters, 19, 145
  distance to, 29, 31
  insects and, 48
  landing, Pioneer, 20
  model, 19, 145
  mountains, 20
  NASA kit, 20
  navigation by, 48
  orbit, 145
  phases, 18–20
  moons and planets, 145
  moth, 52, 58–59
Mother Nature, 167–168
motion

Index

airfoil, 196–197
car, 160–161
escape velocity, from Earth, 143–144
kite, 195–197
mechanical, 160–161
parachute, 198–199
rotation (orbit), 139–140, 145, 152–153
sailboat, 195–197
tacking, 197
whirligig, 198–199
motor, electromagnetic, 131
motor, gears, 160–161
mountains, 20, 26, 41, 68–70, 74; see also volcano
Mount Everest, 41
movement, mechanical of gears, 119, 160–161
lever and fulcrum, 161
pulley and wheel, 161
rotary, 161
mushroom art, 173

N

NASA Moon kit, 20
Natural Laws, 134–161
natural preservatives, 75
nectar, 53–54
nectarine, 170
new moon, 19
newspaper recycled, 81–82
nickel, 115, 155
nitrogen, 172
nocturnal insects trap, 47–48
nonmagnetic metals, 154–155
Northern Hemisphere, 140
North Pole, 10, 11, 12–13, 112, 113, 139–140, 158
Earth's rotation at, 152–153
geographic vs. magnetic, 12–13
north pole, magnet, 12–13, 146–147
north-seeking pole, 13
nucleus (ice), 67
nutrients, 63, 65, 90–110, 166
nutrient tubes, plant, 166
nylon stocking trick, 130

O

oak trees, 79
object, defining
clinograph, 182–184
distance, 27–29, 182–184
echo-location, 40
height, calculating, 182–184
kite-sighter, 182–184
position change, 27–29, 30–31
stationary, 182
ocean
depth indicator, 189–190
echo-location, 40
floor mapping, 40–41
Marianas Trench, 41
mountains, 41

Pacific, 41
plastic trash, 78
sea creatures, 78, 93
temperature, 93
oil
aromatic/volatile (essential), plant, 61
and fat, test for, 92–93
with iron filings in suspension, 146–147
and magnetism, 147
as pollutant, 46
toxic plant, 165
urushiol (poison ivy and oak), 165
and water mixture, 83
onions, 171
orange, 94–95, 132
orange juice, 90–91
orbit, 144, 145
orb web, 55
organisms, 108, 163–164
osmosis, 63, 65
otters, 46, 93
overcrowding, city, 200
oxygen, 85–86, 110, 163–164, 167
deprivation, 163–164
ripening food, 167–168
scale, 85–86

P

Pacific Ocean, 41
pahoehoe, 70
paint, milk, 77
Painted Desert, 25
palm trees, 79
paper, brown, test for fats, 92–93
paper-making, 81–82
papier-mache, 26, 68
parachutes, 198–199
parallax effect, 183
parallax principle, 27–29
parallax shift, 30–31
parallel venation, 79
parasites, soil, 45
particles, abrasion, 148
particles, charged, 124, 129, 130
patina, 123
pattern recognition, insects', 48
pavement, cracked, 26
peach, 170
peas, 99–100
pendulum, Foucault's, 139–140
perfume-making, 61–62
periscope, box, 38–39
permeability, 63, 64, 65, 180
pests, 45, 171
pets, caution, 165
pet water dish, self-filling, 141–142
petroleum products, 77
petroleum plastic, 78
pheromone trail, 51–52
pheromones, 51–52
phloem, 166
phosphates, 163–164
photograph experiment steps, 8
photosynthesis, 104, 105, 163, 174–175
pigment, green (chlorophyll), 163
pink sky of Mars, 17
pings, counting, 40–41
Pioneer Moon landing

(1969), 20
pivot, 28
planet, 145
plank bridge, 135
plant
algae and, 163–164
chlorophyll, 163
circulatory system, 166
cuttings, 171
drought-resistant, 172
embryo, 173
energy generation, 174–175
energy storage, 104, 105
fertilizer, 171
field guide, 165, 178
flowers, 45, 48, 58–59, 79
food manufacture within, 166, 174–175
as food source, 99
freshwater, 163–164
fungi, 170, 173
grafting, 169–170
hybrids, 170
insect fertilization/preference, 48, 58–59
leaves, 79–80, 166
low-water, 172
material, plastics, 77
microscopic, 173
monocots and dicots, 77–78
nutrients for, 171–172
nutrient tubes, 166
oxygen deprivation, 163–164
phloem/xylem, 166
"photograph," 173
photosynthesis in, 104, 105, 163, 174, 175
pods, seed, 165
poisonous, 165, 178
potting, 172
primitive, 173
pulpy tissue, 79
reproduction, 173
roots, 104, 105, 171–172
seeds in fruits and vegetables, 170
seed hooks/barbs, 165
seeds, seedlings, and sprouts, 165, 171–172
spores, 173
starch content, 104
stalk, stem, or trunk, 166, 178
tree, 79–82, 150–151, 166, 174–175, 177–179
variety, 170
venation of leaves, 79–80, 178–179
xylem/phloem, 166
plaster of paris, 148
plastic(s)
animal and plant, 77
bag generating electricity, 129, 130
biodegradable, 78
creamy, 77–78
decomposition, 77, 78
electrical charge in comb, 116–117
fertilizer, 78
homemade, 77–78
plating metal, 155, 157
plum, 170

plunger, 137–138
pods, seed, 165
poison ivy, 165, 178
poison oak, 165, 178
polarity indicator, 132–133
poles, magnetic, 152; see also compass
attraction/repulsion, 146–147, 158–159
battery, 132–133
Earth's North and South, 10–11, 12–13
geographic, 10–13, 152–153
illustrating, 146–147
iron and, 10, 11, 146–147, 158–159
of magnet, 10–11, 146–147
magnetic, Earth and, 10–13
magnetite and, 11
polishing silver, 122–123
polishing stones, 148–149
pollen fertilizing plants, 59
pollen lines, 59
pollution
algae growth, 163–164
animals, and, 46, 163–164
"blue," 16
birds and, 46
detergents and soap in water, 46, 163–164
eutrophication, 163–164
feathers, action on, 46
garbage, super-compressed, 200
groundwater, 46
lakes and ponds, 46
oil on water, 46
oxygen deprivation, 163–164
phophates, 163–164
plant growth, 163–164
types in sky, 16–17
of water, 46, 163–164
wool and, 46
polyethylene, 129, 130
pomato, 169–170
pop can, 154–155
porosity, rock, 66–67
Portland, Oregon, 140
potato, 65, 103–104, 132–133, 169–170, 171
potato polarity indicator, 132–133
potato chip, 92–93
position line (star movement), 28–31
power, electric consumption, 118–119
praying mantis, 48, 58
preservative, natural, 75–76
drying as, 75
lemon juice, 75
nontoxic, 76
salt, 75–76
vinegar, 75
press, paper, 81–82
pressure
air, 196, 199
atmospheric, 138
water, 189–190
proboscis, 53–54
project step-by-step, 5–7

properties, ferromagnetic, 156–157
protein(s), 110
animal, 99–100
breakdown, 101
build/repair body, 99
caesin, 78
combining, 100
complete, 100
cooking, 101–102
egg, 100, 101
heat and, 100, 101–102
testing foods for, 99–100
vegetable, 99–100
protons, 120, 129, 130
protractor, 12, 27–28, 32, 143–144, 150
Proxima Centauri, 22
psi, 138
puffed wheat, jumping, 120–121
pulleys, 161
Puzzling Plant Projects, 162–180

Q

quartz, 149
quill, feather, 46
quinine water, 97
quotient, 138, 174

R

raft, floating city, 200
rare-earth metals, 34–35
record (CD) player, 131
recycling paper, 81–82
red blood cells, 107
refraction, light, 21, 22
refraction patterns, 23, 24
refrigeration, 75
report writing, 6–7
repulsion, electrical, 67, 116–117, 120–121, 124, 129, 130, 146–147
repulsion (force), 158–159
reservoirs, natural groundwater, 46
rheostat, 127–128
rice, 100, 104, 105
rings, tree, 166
ripening fruit, 96, 167–168
ripening hormone (ethylene), 167
rocket, 143–144
rock(s)
flowing, 71–72
metamorphic, 87
molten, 71–72
porous, 66–67
quarry, 87
smoothness, 148–149
tumbler, 148–149
roots, plant, 104, 105, 171–172
rotation of Earth, 12–13, 31, 139–140, 152–153
rotation, gear, 160–161
rotor-vortex microbursts, 74
rotor, whirligig, 198–199
rubberized bones, 110
rubberized egg, 63, 64
rubbing alcohol, 61, 83
rubbings, leaf, 178–179
Russian hydroponics, 171
rust, 85–86, 122–123
oxygen scale, 85–86

Index

S

sailboat movement, 195–197
Saint Petersburg, Russia, pendulum, 140
salt(s), 65, 75, 98, 112, 113, 114
  as electrolytes, 113
salt, Epsom, 66
saltwater
  as conductor, 112–113
  nonsaltwater and, 65
  and plastics, 78
sand
  grain size, landfill, 25
  particles, nonmagnetic, 36–37
  as weight, 135–136, 139–140
sandpaper, 36, 85, 148, 191–192
sandstone buttes and erosion, 25, 148
satellite, space, 144
scale, oxygen, 85–86
science-fair project, 5–8
  analyzing, 5
  bacteria and fungi, prohibition, 6
  building, 7–8
  cautions and safety, 6; *see also specific project*
  chemicals, 6
  choosing, 6
  costs, 6–8
  display tips, 5, 7–8
  experiment, 5
  format, 6
  getting started, 6
  hazardous or toxic materials, 6
  helpers, 7
  insects, 6
  journal, 6
  judges, 6, 7–8
  library work, 6
  live animals, 6
  measurements, 6
  note-taking, 6
  questions, asking, 6
  report, 6–8
  research, 5–6
  rules and regulations, 5–6, 75–76
  sources, 7
  stand, small, 177, 189
scientific method, 5
  hypothesis, 5, 6
scroll, ancient Greek, 161
sculptures, 87–88
sea creatures, 78, 93
seals, 93
seed(s)
  barbs and hooks, 165
  fruits vs. vegetables, 104, 170
  sprouts, 165, 171–172
selective scattering (light), 17
semipermeable, 63, 64, 65, 180
shale, 149
sheet erosion, 25
ship
  buoyancy, 200
  depth gauges, 189
  echo-sounding devices, 40–41
  hydrodynamic hull design, 191–194

measuring ocean depth, 40–41, 189–190
ocean liner, 200
periscope, submarine, 38
speed, 191
weight, 200
shock absorbers in cars, 158–159
shrimp shells, 78
sighter, kite, 182–184
sighter, paper clip-and-straw, 27–28
silver tarnish, 122–123
sinking vs. floating, 200
sky's blueness, 16–17
skywatching, 16–17
Smithsonian Institution, 140
smog, 16
snout, flexible, 53
soap as emulsifier, 83
soap-contaminated water, 46
sock garden, 165
socks, seedy, 165
soda-pop can, 154–155
sodium chloride, 113; *see also salt*
soil
  composition, 34–35
  erosion, 25, 150–151
  fertilizer for, 78
  irrigation, 171–172
  landfill, biodegradable materials, 77–78
  movement, 150–151
  parasites trap, 45
  substitute, 171–172
  water content, 171–172, 174
  watering, 171–172
  xeriscaping and, 172
solar eclipse, 19
solar panels and tree, 174–175
solid/liquid, 71–72, 92
solid surface, Earth's, 153
solution (chemical), 99, 180
sound, speed, and echo, 40–41
sour taste, 98
Southern Hemisphere, 140
South Pole, 10–11, 12–13, 112, 113, 158
  Earth's rotation at, 152–153
  geographic vs. magnetic, 10–13
  south pole of magnet, 12–13, 146–147
south-seeking pole, 13
space station, 144
space travel, 143–144
sparrows, 57
species, 79, 179
speed
  of Earth's rotation, 152–153
  of echo, 40–41
  escape velocity, 143–144
  of kite, 196
  of light, 22
  of parachute, 199
  of rocket, 143–144

of ship, design, 191
of sound (echo-location), 40–41
of toys, electric, 128
speedometer, 131
sperm whales, 62
sphere, 145, 152
spiders, 55–56
  black widow, 55
  "garden," 55
  in outerspace, 56
  poisonous and orb web, 55
spiderwebs, 55–56
  foundation thread, 55
  types, 55
  spun in outerspace, 56
spinach, iron content, 107–108
spinnerets, 56
spinning, 131
split a stone, 26
sponge, growing seedlings in, 171–172
spores, plant, 173
spring
  bouncing, 112–113
  floating-magnet, 158–159
  forming closed electrical circuit, 112–113
  as magnet/magnet as, 112–113, 158–159
sprouts, 172
stalactites, 66–67
stalagmites, 66–67
stand (cardboard support), 177, 189
starch
  breakdown in photosynthesis, 105
  dextrin in, 103
  in greenhouse gas, 168
  heating, 103–104, 106
  iodine test, 105, 106, 180
  in maltose, 95
  molecule, 103–104, 105, 180
  plant energy store, 104
  in ripening fruit or vegetable, 167–168
  solution, 180
  to strengthen paper, 82
  water, 180
starlight, viewing, 21–22, 23, 24
stars, distance, 22
static electricity, 116, 120–121, 124, 130
  balloon, 121
  comb demonstration, 120–121
  flash dancers, 124
  puffed wheat jumping, 120–121
  wool and, 121, 124
stationary object, 182
steel (iron alloy), 155
steel, stainless, 85
stellar parallax, 5
stellar refraction, 23–24
stern (rear), boat, 193
stickiness of spores, 173
stimulant, 167
stocking trick, 130
stomata, 179
stone
  buildings, 88
  papier-mache, 26
  porous, 67

smoothness, 148–149
-splitting, 26
tumbler for polishing, 148–149
wall, 151
Stratos, 161
strontium, 35
Styrofoam ball on spring, 112–113
submarine periscope, 38
sucking-salt solution, 65
sucrose, 95
suffocation, pond animals and plants, 163–164
sugar(s)
  complex, 95
  conversion, 96, 103, 106
  digestible, 105
  in foods, 94–95
  heating, 95
  photosynthesis and, 174–175
  from starch, 96, 103, 105, 106, 167–168, 174–175
  simple, 95
  testing, 94–95
  types, 95, 103
  water, 94–95
sulfuric-acid solution, 88
Sun, 17, 19, 48, 104, 105
  Earth's rotation and, 139–140
  eclipse of, 19
  energy, 105, 174–175
  insect navigation by, 48
  photosynthesis and, 104, 105, 174–175
  sunlight, 16, 17, 174–175
  collecting, 174–175
  reflected in Moon, 20
  tree as solar panel, 174–175
sun position, 17
super-cooled liquid, 72
surface area, 138, 174–175
surface, curved, 196
surveyors, 30
suspension, 147
swans, 46
sweeteners, 95, 97–98
swimming and fat, 93
swimming spore cases, 173
swing, 139–140

T

table salt (sodium chloride), 113
tacking (sailboat), 197
tail of airplane, 185–188, 195–197
tail of kite, 195–196
tannic acid, 107
tarnish, removing silver, 122–123
taste, 97–98, 167–168
tea, 107, 109
telephone pole lean, 150–151
telescope, 23
temperature, 173
terminal indicators, battery, 132
test standard, for detecting vitamin C, 90
third force, 161
thundercloud, 116
ticks, caution, 165

tides, Earth, 153
time interval measurement, 41
tin, 155
tissue (body) repair, 90
titanium ore, 35
titration, 90–91
toast, 106
Tokyo, Japan, overcrowding, 200
tomato juice, 90–91, 113
  as electrolyte, 113
tomato plant, 45, 94–95, 167–168, 169–170
  as fruit, 170
  graft, potato, 169–170
  hot-house, 168
  hybrid, 170
  ripening fruit, 167–168
tomato sauce, 94–95
Tom Collins mix (quinine water), 97
tool, 182; *see also specific tools*
tornado, horizontal, 74
tornado vortex, 73–74
toxic plant oil, 165
trail, ant, 51–52
trains, high-speed, 158–159
transfer of electrical charge; *see electricity*
tree
  bark, 166
  circulation, 166
  evolution, 166
  identifying, 178–179
  leaves of,
    count, 174–175
    rubbings, 178–179
    shape, 176–177
    skeletonized, 79–80
    venation, 79, 166, 179
  lean on hill, 150–151
  Michigan species, 179
  papermaking, 81–82
  photosynthesis, 174–175
  and poisonous plants, 165
  rings, 166
  saving, 82
  shade, 79
  soft-pulp, 81–82
  solar machine, 174–175
  species, 177, 178–179
  trunk, 151, 166
  xylem/phloem, 166
trial-and-error, 135
triangle
  formula, 32
  parachute design, 199
  right triangle, 32–33
  sides of, 32–33
  spiderweb (orb), 55–56
triangular-truss bridge, 136
triangulation, 27–29, 183
  to calculate distance, 27–33, 182–184
  to calculate height, 32–33, 182
  and circle, 29
  clinograph, 182–184
  instrument, 27–29
  and parallax principle, 27–33
trigonometry, 27, 29, 32, trunk, elephant, 53

truss bridge, 136
tube(s), nutrient, 166
tubers, plant, 171
tumbler, rock, 148–149
turgidity, 65
twinkling starlight, 21–22

U

ultra-high spectrum colors, 48
ultraviolet, 48, 59
Up, Down, All Around, 9–41
urinalysis, 94
urushiol, 165
U.S. Geological Survey, 153

V

vaccination, 170
vacuum
  atmospheric pressure and, 137–138
  ball experiment, 138
  creating, 137–138
  motor for, 131
vegetables
  iron content, 109
  ripening, 167–168
  seeds (vs. fruit), 170
  starch content, 104, 105, 168
  sugars in, 168
vegetarian diet, 100
veins, 79
velocity, escape, 143–144
venation, leaves, 79, 178–179
  patterns, 79, 178–179
Vinci, da, Leonardo, 199
vinegar, 10, 63, 68, 75, 77, 83, 87, 94, 110, 113, 114
  as electrolyte, 113, 114–115
  and salt solution, 114–115
violet, 48
viruses, vaccines, 170
viscosity, 71–72
visual diagnostic strips, 94
vitamin C, measuring, 90–91
vitamins in cell growth and repair, 110
vitamins in vegetables and fruits, 170
volcano
  atomic explosion, resemblance, 69
  Crater Lake, 69
  erupting, 68–69, 70
  Hawaii, 70
  lava-flow, 68–69, 70
  mountains and, 68, 69
  pressure, 69
  shooting, 68–69
volume (noise), rheostat reducing, 128
von Guericke, Otto, 138
vortex, 73–74

W

wake (water), 193
walruses, 93
waning moon, 19
warfare, Chinese with kites, 197
Washington, D.C., 140
wastes, cell metabolism, 63, 65, 180
water
  abrasion, 148
  algae growth, 163–164
  boiling point, 85
  carbonated, 97
  current, 194
  depth indicator, 40–41, 189–190
  deuterium and, 85–86
  displacement of, 200
  droplets, 67
  echo-location, ocean, 40–41
  erosion, soil/ sand, 25
  flow pattern, 176, 193
  freshwater pollution, 163–164
  "heavy," 85–86
  hydrogen removed, 85–86
  and hydroponic gardening, 171–172
  as ice, expansion, 26
  irrigation, 171–172
  leaf repelling, 176–177
  mineral-rich, 66–67
  molecules, 72
  and oil mixture, 83
  oxygen reduction in, 163–164
  phophates, 163–164
  plants and animals life, 163–164
  pollution, 163–164
  pressure, 189–190
  quinine, 97
  sluice, 191–192
  sugar, 94–95
  stalactite and stalagmite formation, 66–67
  temperature, 93
  turbulence, 21
  wake, 193
  weight of, 85, 176–177
  xeriscaping, 171–172
water dish, self-filling, 141–142
water sluice, 191–194
watt/ watt hours, 119
wavelength, 16–17, 59
waxing (crescent/ gibbous) moon, 19
weatherproofing, 77
web preservation, 55–56
weevils, 45
weight
  of air, 137–138
  -less environment, 56
  loss, 93
  of oxygen, rust, 85–86
  sand, 135–136, 139–140
  support for, 135–136
  of water, 176–177
whale, ambergris, 62
wheel mechanisms, 161
whipped cream, fat test, 92–93
whirligigs, 198–199
whirlpool, 73–74
whirlwind, 74
wind, 73–74, 148, 187–188, 196–197
wind baffle, 185–187, 191
wind test, 187–188
wing, 185–188, 196–197, 199
wire coil behaving like magnet, 112–113
wood pulp, 81–82
wool, static electricity, 121, 124
Wright brothers, 197

X

xeriscaping, 172
xylem and phloem, 166

Y

yellow, 59
"yellow-jacket" bees, 59
yogurt, fat test, 92–93
yttrium-iron garnet, 35

Z

zigzag (tacking) movement of sailboat, 197
zinc, 35, 122
zone of partial melting, Earth's, 71–72

About the Author

Glen Vecchione is an author, illustrator, composer, and lyricist, living in southern California. He also designs web pages for a Fortune 30 communications company.

He has written musical scores for animated science films, including *Matter & Anti-Matter* and *The Metric Film*, which won Golden Cindys. He has also written themes and jingles for CBS television spots, including their fall lineup commercials and other on-air promotions.

He has composed musical scores for theater pieces and a few Off-Broadway shows, including *The Legend of Frankie and Johnny*, which went on national tour with the Nat Horne Dance Theatre. With Quiet Zone Theatre, a San Diego–based company of physically challenged actors and dancers, he adapted the Dr. Seuss book *Bartholomew & the Oobleck* for American Sign Language.

His poetry has been published in *Southern Poetry Review*, *Indiana Review*, and *Jacaranda Review*, and he has won many poetry awards.

Mr. Vecchione has written and illustrated several books for children and adults, including the Sterling books *The World's Best Street & Yard Games* (1989), *100 Amazing Make-It-Yourself Science Fair Projects* (1994), *Magnet Science* (1995), *Challenging Math Puzzles* (1997), and *The Little Giant Book of Kids' Games* (1999).

Excerpts from his science books have appeared in short series in national magazines and newspapers, like the *Cincinnati Enquirer* and *The Detroit Free Press*.

In his spare time, he likes to tinker with and invent new gadgets. His home is filled with all kinds of projects and experiments in various stages of completion.